# ORGANICS IN THE
# GLOBAL FOOD CHAIN

# Organics in the Global Food Chain

Edited: Bruno Mascitelli & Antonio Lobo

Connor Court Publishing
Ballarat

Published in 2013 by Connor Court Publishing Pty Ltd

Copyright © Bruno Mascitelli and Antonio Lobo 2013

ALL RIGHTS RESERVED. This book contains material protected under International and Federal Copyright Laws and Treaties. Any unauthorised reprint or use of this material is prohibited. No part of this book may be reproduced or transmitted in any form or by any means, electronic or mechanical, including photocopying, recording, or by any information storage and retrieval system without express written permission from the publisher.

PO Box 224W
Ballarat VIC 3350
sales@connorcourt.com
www.connorcourt.com

ISBN: 9781922168856 (pbk.)

Cover design by M. Giordano

Photo: Don and Elaine Murray from "Nature's Haven". Photographer Margaret Murray. Used with permission courtesy of Australian Organics.

Printed in Australia

# CONTENTS

| | |
|---|---|
| Acknowledgements | vii |
| Abbreviations | ix |
| List of Tables and Figures | xi |
| Preface: *Liz Clay* | xiii |
| 1. The many sides and shades of the organic industry – *Bruno Mascitelli and Antonio Lobo* | 1 |
| 2. The science of organic food – *Bruno Mascitelli* | 17 |
| 3. A history of the organic agriculture movement in Australia – *John Paull* | 37 |
| 4. Profiling the Australian organic market – *Andy Monk* | 61 |
| 5. Consumer preferences for organic food and LOHAS – *Antonio Lobo* | 77 |
| 6. Organic food supply chains: Challenges and prospects – *Chandana Hewege* | 97 |
| 7. Global trends in organics – *Andre Leu* | 119 |
| 8. A case study of trends in the Chinese organic food market – *Jue Chen and Barry O'Mahony* | 153 |
| 9. Contemporary issues related to food security – *Mark Gibson* | 175 |
| 10. Food sovereignty in a post-organic era – *Alana Mann* | 201 |
| References | 231 |
| Author Biographies | 271 |
| Index | 275 |

# Acknowledgements

The idea for this book, as with many other things in life, occurred as a result of serendipitous events which were left-field from our normal day to day activities as academics. It was an idea, that once proposed, seemed like a 'no-brainer'. One of our students decided to undertake research on organic food in China and the two editors of this study were the supervisors. At first we felt distant from this theme but within little time we became ultimately hooked on the topic. Like the help we received for the idea of the book, we also benefited from many suggestions and ideas on how best to produce a stronger 'academic' focus on food and especially organic food. We as editors were learning more about organic food but there were already experts in the field who were best suited to meet this 'literature gap'. Therefore it was natural that they, with the knowledge, be the authors of the chapters. We thank our authors for their expertise and timely contribution to this book. As we have begun to understand some of the facets of this unusual industry with idiosyncratic stakeholders we have realised there are many aspects of this industry which go beyond the realm of this book. Without doubt we will have certainly missed areas worthy of research and study and we apologise from the outset for this.

This industry has a multifaceted nature which is a mixture of diverse philosophical views on living and quality of life, a convinced concern about pesticides and chemicals and other facets such as the role of the lifestyle driver for organic consumption. In the case of China, as our readers will see in the relevant chapter, the study demonstrates a clear concern about food safety for its consumers especially of food and dairy products. We have sought to cover as many of these influences as possible and are thankful to the authors for their ability to capture their topics in a concise and timely fashion. Others who have helped along the way include researchers and students and these include Dr Liyanage Chamila Pereira. We are most thankful for the understanding and astuteness of Connor Court Publishers who understood the value of this project and supported it from the beginning.

While we as editors have sought to ensure the correctness and fairness of the commentary of the authors, the ultimate responsibility for the content presented in each chapter is that of the authors.

**The Editors, 1 September 2013**

# Abbreviations

| | |
|---|---|
| ABC | Australian Broadcasting Commission |
| ACO | Australian Certified Organic |
| ACP | Agriculture and Consumer Protection |
| AEC | Agricultural Experimental Circle |
| AOFGS | Australian Organic Farming and Gardening Society |
| AUSFTA | Australia-United States Free Trade Agreement |
| BFA | Biological Farmers of Australia |
| BSE | Bovine Spongiform Encephalopathy |
| CFSC | Conference of the Community Food Security Coalition |
| CSIRO | Commonwealth Scientific and Industrial Research Organisation |
| EU | European Union |
| Food miles | Distance which food travels between production and consumption |
| FAO | Food and Agriculture organisation of the United Nations |
| FTA | Free Trade Agreement |
| G8 | Largest eight economies in the world |
| GATT | General Agreement of Tariffs and Trade |
| GDP | Gross domestic product |
| GM food | Genetically modified food |
| GMO | Genetically modified organism |
| IAEA | International Atomic Energy Agency |
| IFOAM | International Federation of Organic Agriculture Movements |
| IFPRI | The International Food Policy Research Institute |

| | |
|---|---|
| IMF | International Monetary Fund |
| LSAT | Living Soil Association of Tasmania |
| LOHAS | Lifestyles of Health and Sustainability |
| NASAA | The National Association for Sustainable Agriculture Australia |
| OA | Organic Australia |
| OFD | Organic Farming Digest |
| OECD | Organisation for Economic Cooperation and Development |
| OGFST | Organic Gardening and Farming Society of Tasmania |
| OTA | Organic Trade Association |
| SASA | Soil Association of South Australia |
| SASAB | Soil Association of South Australia Branch |
| TNC | Transnational Companies |
| UK | United Kingdom |
| UN | United Nations |
| USA | United States of America |
| USDA | United States Department of Agriculture |
| WHO | World Health Organisation |
| WTO | World Trade Organisation |

# List of Tables and Figures

## Tables

| | | |
|---|---|---|
| Table 1.1 | Definitions of organic food and farming | 8 |
| Table 2.1 | 13 reasons why organic was better than conventional food | 20 |
| Table 2.2 | Genetically modified crops (million hectares) | 34 |
| Table 3.1 | Australian members of the Agricultural Experimental Circle of Anthroposophical Farmers and Gardeners | 41 |
| Table 3.2 | Milestones of the development of organic agriculture in Australia | 45 |
| Table 6.1 | Organic supply chain difficulties | 104 |
| Table 7.1 | Improved Organic Productivity: Mean yield of rice, 2007 (kg/ha), n=840 | 150 |
| Table 7.2: | Improved income from organic farm: Net agricultural income per hectare, 2007 (Pesos) | 150 |
| Table 7.3 | Improved income from organic farm: Annual Balance of Income and Expenditure per Household, 2007 (in Pesos), n=840 | 150 |
| Table 8.1 | Profile of the respondents | 165 |
| Table 9.1 | Issues of Food Security Consideration | 181 |

## Figures

| | | |
|---|---|---|
| Figure 3.1 | World map of organic agriculture with countries proportioned according to the tally of certified organic agriculture hectares | 39 |
| Figure 4.1 | Australian Retail Sales Growth 1990-2012 | 70 |
| Figure 4.2 | Australian Farm Gate Sales Growth | 71 |

| | | |
|---|---|---|
| Figure 5.1 | Influences of LOHAS – Values, Worldview, Lifestyle | 88 |
| Figure 7.1 | Global Organic Market Access Logo | 127 |
| Figure 7.2 | The IFOAM Family of Standards | 129 |
| Figure 7.3 | Examples of IFOAM Recognised PGS Logo | 131 |
| Figure 7.4 | Countries with PGS Systems | 132 |
| Figure 7.5 | Impact of using compost – Grain yields from over 900 samples from farmers' fields over 7 years | 138 |
| Figure 7.6 | Frequency of relative yields of organic vs. conventional | 140 |
| Figure 7.7 | Diagram of a Push Pull system in Maize | 142 |
| Figure 7.8 | Silver Leaf Desmodium growing in Maize in a Push Pull System | 143 |
| Figure 7.9 | Kenyan Farmer standing next to Napier grass that has been progressively cut and fed to a cow | 144 |
| Figure 8.1 | Global market for organic food and drink: Market growth 2000-2010 | 154 |
| Figure 8.2 | Labels of food categories in China | 163 |
| Figure 9.1 | Levels of Food Security | 187 |
| Figure 9.2 | Historical and Projected Population Trends | 190 |
| Figure 9.3 | Food Availability (per person per day) | 193 |

# Preface

*Liz Clay*

Organic agriculture means different things to different people. For many, it is the notion that food grown in rich, biologically active soil without the use of artificial fertilizers and harmful chemicals is good for you and the environment. For some, organic food is not just about the end product but a part of a connected and ethical approach to the whole food and fibre system. Indeed for this group of people, organic agriculture is an entire process as compared to food just to be consumed. Respectful relationships with all involved including farmers, workers and consumers, and the respectful treatment of animals and of the earth are expected to be embodied in organic food. Having stated this, there are many others who do not support any of these notions unconvinced that organic provides any particular benefits particularly when there is a higher price tag on organic products.

What is agreed, however, by the global organic movement are the principles of organic agriculture, which are health, ecology, fairness and care. These principles outline the great scope of organic agriculture and have been adopted by many national governments. They provide guidance to define organic agriculture including in the development of standards for organic products and therefore are reflected in certified organic food available across the world. Its true being involved either as a producer and/or consumer of organic food our actions are voluntary – no one is forcing the increasing number of farmers across the world to grow organically, no one is compelling people to choose organic food over other. Yet the production and consumption of organic food is and has been growing steadily across the world for the last twenty years and Australia is no exception.

It is also true that many governments have established programs to develop organic agriculture in their countries to capitalize on the growing market place. Indeed a number of countries have targets for conversion to organic production. The Australian government over the last twenty years have supported the organic sector through providing a well-respected third party certification and accreditation program for organic product exports which has enabled a vibrant export sector. However, scant support has been provided through public programs to support the development of the organic sector domestically.

Pulling together a range of contributions by writers both well-known to the organic sector and independent observers, this book offers a solid basis to understand some of the drivers and performance of the organic market place particularly in Australia. In a world where our vision is filtered by dominant world views and values favouring single cultural ways of doing and being, organic agriculture in all its diversity invites alternative, multidisciplinary ways of understanding and thinking that places our human activity sustainably within nature.

The topics covered in this single volume are not comprehensive, but they do provide a sample of the diversity of what organic covers. The reach of organic is greater than agriculture, well-illustrated by the work of the International Federation of Organic Agriculture Movements (IFOAM) as it communicates the multiple benefits of organic agriculture globally. IFOAM, (lead by Australian farmer Andre Leu, also a contributor in this book) is an active participant in global discussions on climate change, food security, food sovereignty, biodiversity and standards for the market place. The fact that organic is multidisciplinary by nature associated with health, food, environmental management, agriculture, biodiversity, genetic diversity, ethics, sustainability, not to mention the economic opportunities and industries that have flowed presents a vast array of topics for engagement. There is no doubt that organic offers new ways of thinking beyond that which is simplistic and reductive and offers new ways of creating economic systems that value people and nature.

This book has established a starting point for these discussions and credit to the authors for having the courage and foresight to delve into this yet under studied discourse. My sincere hope is that the profound knowledge that organic agriculture represents, current and untapped, becomes a part of the necessary conversation as we grapple with the most serious of our challenges – our own survival.

In Australia understanding of and support for organic agriculture is largely focussed on food and the market place. The scanty research that has taken place has been related to markets and consumer information – critically important to popularize organic food. Much of this research is offered within the chapters of this book. However, more recently, I have observed a growing reference to organic agriculture as various disciplines from the health sector, environment, community development, agriculture and business examine emerging issues.

In health, concerns about how food is produced, has brought into question the safety of using genetically modified organisms and pesticides. Nutritionism, a term coined to describe a reductive bias to assessing food quality where food is judged by some of its component parts disconnected from the food as whole, is now critiqued as simplistic, harmful, promising unrealistic expectations about potential benefits and further, distracts from appreciating food's important role in connecting cultures, traditions and communities.

In agriculture over the last five years, a response to climate change has witnessed a changing consciousness in how we should farm. During that time a movement that recognises the role of soil as a living thing providing ecosystem services for production systems and causing farmers to change their existing farming practices, has been rippling across the country side. That agriculture has existed in this country without acknowledging the fundamental role of soil biota in growing food and fibre has bemused organic and biological farmers and is the long established point of difference between organic and conventional approaches to farming. Now we are witnessing the adoption of organic agricultural practices by mainstream agriculture.

On another front, whilst supermarkets remain the main place of purchase for organic food, a growing number of discerning consumers are striking relationships directly with farmers buying organic produce at such places as farmers markets, through box schemes and community supported agriculture programs. Many more ideas to access food directly through food hubs are emerging with the help of new technologies. We are experiencing an emerging food culture with consumers wanting to be more involved with their food, know who grows it, where it is grown and how is it grown. As an organic farmer, this heralds an exciting time for organic agriculture. A time to honour and celebrate together the most generous giving nature of our Earth! Organic is a great story and I hope this book sparks interest in the Organic Agriculture Movement in Australia, so volumes and conversations will follow.

**Melbourne, August 2013**

# 1
# The many sides and shades of the organic industry

*Bruno Mascitelli and Antonio Lobo*

## Introduction

Until the early 20th century, food production and agricultural practices were considered to be *de facto* organic. Hence there was no necessity of specially using the term 'organic food' as all food produced was generally so. The early 1900s saw the emergence of the use of fertilisers and chemicals in agriculture and farming with the main intention being that of boosting agricultural production. This was supported by greater levels of mechanisation and efficiencies thus resulting in the phenomenon of 'agricultural industrialisation'. Later in the 20th century agricultural procedures were further refined thus producing the 1960s and 1970s 'Green Revolution'. The organic farming movement emerged and gained momentum in the 1940s in a number of developed economies, essentially to combat the perceived ill effects of agricultural industrialisation. Organic farming practices, amongst other things saw the need to return to the use of technology with limited side effects, the return of traditional farming methods and consideration for ecological balance and eco-systems disruption, all of which would contribute towards natural biological processes. In many respects it was a response to the mass production of agriculture and the synthetic chemicals which became the backbone of the modern horticultural industry. On the surface, consumers in the twentieth century benefited from the 'industrialisation of agriculture', through the economies of scale, specialised production, new technologies and new efficient food production systems. But behind this reality of boosted yields, availability of agricultural products, and economically accessible agricultural products, not all was what it seemed. The organic story offers another view of where humanity stands in terms of food production and consumption. This we will hear throughout this study.

In general, 'organic food' is perceived by many as being food that is produced without using synthetic pesticides and chemical fertilisers. Because of the non-use of pesticides and chemical fertilisers, this term can also include non-food products such as beauty products, beverages as well as processes such as 'organic farming', and 'organic agriculture'. As such readers will need to appreciate the wider interpretation of this term.

This chapter introduces and covers a range of features of the organic industry. The section that follows provides an overview of the historical evolution of organic food and agriculture. We then turn to the definitions and terms associated with the characteristics of organic food. The section that follows discusses the antecedents of organic food consumption as well as general trends in organic food consumption in the global and Australian context. To this end, the purpose of this chapter is to discuss antecedents, drivers and trends of organic food consumption practices among all stakeholders including consumers.

Though Albert Howard is often characterized as the father of organic agriculture, and he was thankful for this recognition, he preferred to acknowledge the pioneering role in the development of 'organic' played by others in history. Keating captured this sentiment in stating that Howard was more appreciative of the role played by:

> … the generations of Eastern peasant farmers, primarily in China but throughout South East Asia who handed down organic principles to become, as Howard's peer F.H. King noted, 'Farmers of Forty Centuries'. These origins and the influence of belief systems that shaped them – specifically Buddhism and Hinduism – help explain why organic agriculture so often seems counter-intuitive to Western agriculturalists. Fortunately, Westerners such as

Howard and King (an American) were not put off by the primitive appearance of Asian agriculture and gleaned its magnificent substance (Keating 2010).

## The politics of the organic industry

Few industries have such a high degree of political association and politicisation associated with it as does the organic (food) industry. As one of the contributors in this study has pointed out, there is the corporatist, the progressive and the radical view of food and its belonging. Food is a key feature of the existence of many societies and its politicisation is but one of the themes which emerges in this collection of essays. This is very much the case also with 'organic'.

No history of organic agriculture would be complete without addressing the most frequently asked question about this subject: Are organic foods healthier for people? Howard attributed the nutritional superiority of organic food to the abundance of mycorrhizal fungi found in biologically active soils sustained by compost, crop rotations and cover crops. These fungi penetrate the fine root hairs of neighboring plants in a mutually beneficial relationship that facilitates nutrient uptake in both. Howard also thought that these characteristics were transmitted through subsequent relationships in the food web, such as livestock grazing on healthy pasture and humans consuming food from organically raised crops and animals. Conversely, Howard postulated that a microbiologically weak soil would yield deficient amino acids and proteins that would invite disease in the plants, livestock and people that subsequently consume them (Keating 2010). This is not a scientific justification for the importance of 'organic' but in examining the industry holistically, as is done in Chapter two. We will discover that the rise of the organic movement and how it engaged with the mass production of conventional food is a matter of debate and not necessarily an accepted one. Have the observations made by Howard stood the test of scientific scrutiny?

## Historical evolution of organic food and agriculture

The terms organic food and organic agriculture are used interchangeably as both of them share common aspects. Put plainly, the growing process of organic food is referred to as organic agriculture. As noted in the previous section, prior to agricultural industrialisation, all food was considered to be grown organically without the usage of synthetic pesticides and chemical fertilisers. However, today, without proper organic certification it is difficult for one to determine if a particular type of food is grown organically or not (Halpin 2004).

The first organised movement responding to agricultural industrialisation commenced in Germany in the 1920s. This movement was centred on resisting the use of chemicals in farming practices. The rise of this understanding of organic food and agriculture also coincides with significant contributions from several key scholars and practitioners throughout the first decades of the twentieth century. Among them, the work of Rudolf Steiner and Albert Howard in the 1920s frequently appeared in the historical records of organic agriculture. According to Hall (2011), Rudolph Steiner during the 1920s in Poland made important references to the development of the concept of 'bio-dynamic farming' – a farming methodology built on the interrelationships of the soil, plants and animals. Furthermore, Steiner's concept of 'bio-dynamic farming' contributed towards the establishment of a formal system of certification and labelling named Demeter, which was later associated with the Demeter symbol. This certification system has become one of the largest organic certification and labelling systems in the world and is currently used by approximately 4,200 producers in 43 countries worldwide.

In 1940, Walter Northbourne introduced the term 'organic' as well as other agricultural practices, such as managing a farm as an 'organic whole' in his book titled *Look to the Land*. In the same year, Jerome Rodale introduced the term 'organic agriculture' and other organic agricultural practices in the USA. Albert Howard, added to these developments by publishing his book titled '*An Agricultural Testament*' in 1943 (Howard 1943; Heckman 2005; Hall 2011). This publication is in hindsight considered one of the seminal works in the organic agricultural movement. Essentially, Howard argued that all farm production should be based on 'the Law of Return' whereby soil fertility is ensured and acknowledging there was significant concern about soil mismanagement which became central to the concept of organic farming.

In 1946, Eve Balfour, one of the founding members of IFOAM, encouraged by Albert Howard, established the Soil Association in the UK (Institute of Food Science and Technology 2009). Eve Balfour, in her own way added to the body of knowledge at the time by publishing '*The Living Soil*', a book that compared organic and non-organic farming methods. Finally in 1962 the extant literature was joined by the historic addition of '*Silent Spring*' by Rachel which further motivated consumers to engage in organic food consumption. In more recent times the fight against agricultural industrialisation took on another feature – that of globalisation. Globalisation innocently became another feature which drove many to organic farming and the search for sustainability in food production and farming. Because the United States played an important role in enhancing the consumer acceptance of organic food, the US Department of Agriculture (USDA) 1980 report and recommendation on organic farming provided an impetus to organic agriculture and gained the attention of many consumers and farmers. By the 1990s a sizeable number of consumers had become interested in organic food, food safety, and environmental wellbeing. In that same decade the US enacted the Production Act which promoted organic farming in the USA and established the official labelling and certification by the USDA which was completed by 2002.

## The slow inexorable rise of the organic industry

There are many aspects of organic which bring into play strong differences of food and as a result lifestyle. There is a morality of food, a lifestyle of organic versus a business, and of course the matter of genetically modified food. In some quarters there is also the desire for the re-connection between the consumer and the farmer.

The contemporary organic food industry which originated in the early 20th century is continually growing worldwide. Sales of organic food and beverages in the USA in 2010, recorded a 7.7 per cent growth rate (Organic Trade Association 2011). According to the annual report of the International Federation of Organic Agriculture movements (IFOAM) – an international organisation which was established with a mission of addressing the complexity of organic agricultural movements – the existing organic market worldwide was worth US $59.1 billion. The transformation of the organic food market from its original niche market status to its current mainstream market status is also highlighted in many academic research publications (e.g., Essoussi and Zahaf 2008; Fearne 2008). According to the more recent publications of IFOAM, 37.2 million hectares of organic agricultural land were available worldwide and countries with the largest ownership of organic agricultural land were Australia, Argentina, and the USA (cited in Heckman 2005, p. 98).

Several factors triggered the rapid growth of the organic food industry. Amongst them, concerns relating to the environment, human and animal wellbeing are widely documented (e.g. Fearne 2008; Kearney 2010; Thøgersen 2010). Although some researchers provide compelling evidence of favourable consumer attitudes toward organic food (e.g. Kihlberg and Risvik 2007; Aschemann et al. 2007), there are others who demonstrate that consumers do not actually purchase organic food (Sahota 2009; Schaack and Willer 2010). This therefore implies that there may be reasons and issues which influence and motivate consumers to purchase organic food.

## The characteristics of organic food

The term 'organic' is rooted in 'bio' which emanates from the Greek word 'bios' meaning life or way of living (Essoussi & Zahaf 2008). The discourse relating to organic food is usually centred on farming or production practices. As such, emphasis on biological, natural, environmentally friendly and limited use of chemicals in production systems are all common terminology used when articulating the characteristics of organic food. In this context the Agriculture and Consumer Protection (ACP) offered a definition which stated:

> Organic is a labelling term that denotes products that have been produced in accordance with organic production standards and certified by a duly constituted certification body or authority. Organic agriculture is based on minimising the use of external inputs, avoiding the use of synthetic fertilisers and pesticides (Agriculture and Consumer Protection 2002).

A further complete definition is offered by the US Department of Agriculture. It says:

> Organic food is produced by farmers who emphasise the use of renewable resources and the conservation of soil and water to enhance environmental quality for future generations. Organic meat, poultry, eggs and dairy products come from animals that are given no antibiotics or growth hormones. Organic food is produced without using most conventional pesticides (United States Department of Agriculture 2005).

While the ACP and USDA have sought to offer holistic definitions, many institutions and scholars of food and organic food have crafted their own slightly distinct meanings and interpretation of organic food. Below are some of these definitions and explanations presented in Table 1.1.

## Table 1.1: Definitions of organic food and farming

| | |
|---|---|
| 1. Organic food is produced without pesticides, herbicides, inorganic fertilisers, antibiotics and growth hormones. Animal welfare is important, and bioengineering and genetically modified foods are not accepted. | Honkanen and others (2006). |
| 2. Organic foods are minimally processed to maintain the integrity of the food without artificial ingredients, preservatives or irradiation. | Green Earth Organics (2008). |
| 3. Organic produce is grown and processed without the use of synthetic chemicals, fertilisers, or GMOs with a focus on environmentally sustainable practices. | Australian Organics (2009). |
| 4. Organic farming is a holistic production management system which promotes and enhances agro-ecosystem health, including biodiversity, biological cycles, and soil biological activity. It emphasizes the use of management practices in preference to the use of off-farm inputs, taking into account that regional conditions require locally adapted systems. This is accomplished by using, where possible, agronomic, biological, and mechanical methods, as opposed to using synthetic materials, to fulfil any specific function within the system. | Food and Agriculture Organisation of the United Nations (FAO), (2013). |
| 5. Organic agriculture is an agricultural production system that promotes environmentally, socially and economically sound production of food and fibres and excludes the use of synthetically compounded fertilisers, pesticides, growth regulators, livestock feed additives and genetically modified organisms. | The International Federation of Organic Agriculture Movement (IFOAM), (2011). |
| 6. Organic farming rejected the chemical-based farming techniques of mainstream agriculture (minimal, definition). | Niggli (2007). |
| 7. i) Growing and processing without chemicals, fertilisers or genetically modified organisms; and<br>ii) Using sustainable farming methods such as crop rotation and organic fertilising. | Faidon and others (2006). |

*Source: The authors, 2013*

With respect to the above definitions, it can be seen that the characteristics of organic food, farming and agriculture are largely centred on zero-chemical use in production systems. According to IFOAM, the four principles which guide organic agricultural practices are: health (health of soil, plants, animals, humans and the planet), ecology (ensuring ecological systems), fairness (harmonious relationships with the environment), and care (wellbeing of current and future generations and the environment).

These principles also dominate the focus of many organic food certification schemes. Nevertheless, contentious issues associated with the definition of organic food have not yet been resolved in the organic food literature.

## General trends in organic food consumption and the Australian scenario

As discussed in the first section of this chapter, organic farming has become one of the fastest growing segments of agriculture across many countries – 82 per cent growth in the period 2006-2008 (Willer et al. 2008). According to IFOAM reports, the growing demand for organic products also offers many opportunities for producers in developing as well as developed countries (IFOAM 2011). However, there has been scant research aimed at systematically comparing organic food consumption across all countries (Daugbjerg and Halpin 2008). It should also be noted as has done the Global Strategic Business Report, that a sizeable number of consumers in the USA, Germany, Great Britain, Denmark, Italy, and Austria tend to eat organic food. Hence, these countries already have small but well-structured markets catering to the production and sale of organic food (Global Industry Analysts 2006).

*Prima facie* Australia has a secure food supply but this is not something it can count on to be long term and nor can this state of play be said to be airtight. Floods, droughts and other natural calamities have unravelled this feeling of security in short periods of time. Moreover there are 135,000 farms and 93 per cent of all produce consumed by Australian consumers is produced in Australia (Loughnan 2012). While conventional food is very much accepted in the Australian food chain, for those looking for alternative arrangements in the way of organic food, without additives, there is a vibrant and growing market of organically, sustainably produced food supplying specialised markets. As a measure of this growing segment, on average the retail organic food market has grown by 20 per cent per year since the 1990s (Loughnan 2012).

According to the Australian organic market report 2010, the organic market in Australia, despite representing only one per cent of the world organic market, was worth A$1.276 billion and was growing rapidly (Monk et al. 2012). It is an industry in which the major supermarkets (e.g., Coles, Woolworths, Aldi and IGA) are allocating more shelf spaces for organic food than they did previously (Mitchell et al. 2010). This can certainly be considered as a favourable development as compared to the slow rate of progress of the Australian organic food industry in general. In this context, some researchers have claimed that although Australia is a major exporter of agricultural products with a high potential of catering to international organic consumer needs, research on the Australian organic food industry is minimal (Chang et al. 2005). Similarly, Lockie (2006) a leading researcher of organic food consumption in Australia, advises caution as there are issues regarding availability of accurate data relating to the consumption of organic food domestically.

Lockie (2006) also highlights the vulnerability of the organic food industry in Australia which on the whole primarily targets the privileged few. Equally problematical is the fact that not all organic operations are sustainable and therefore much progress is required to have a sustainable food network. This was further supported by Halpin and Daugbjerg (2008) who, in their examination of the organic farming industry in Australia, argued that the lack of networking and interaction between the state government and the organic industry in Australia had hampered the growth of this sector.

A decade ago, a focus group study revealed that the most influential factors inhibiting organic food consumption in Australia were cost, inconvenience and unavailability (Lockie et al. 2002). These researchers also reported that similar to the findings in other developed countries, Australian consumers emphasised the importance of organic certification systems. However, only a small percentage of them clearly understood the meaning and characteristics of the different certification and labelling systems in use.

With regards to the factors that drive organic food consumption among Australians, researchers report that Australian consumers are untrusting about the assertions that organic food provides any special benefits. At the same time they perceive organic foods as being relatively more expensive than conventional foods (Bhaskaran et al. 2006). Additionally, it has been reported that demographic differences can be used to explain organic consumer behaviour. More specifically, 44.1 per cent of women respondents claim to have consumed certified organic foods compared to only 33.8 per cent of men (Lockie et al. 2002). Further, the level of education is positively associated with organic food consumption among Australian consumers (Lockie et al. 2002). These researchers also found no differences between organic and non-organic consumers in relation to price, sensory appeal, convenience, familiarity, and religion. Confirming most of the above findings, another study by Lea and Worsley (2005) which had 500 respondents, reported that the majority of Australians believe that organic food is healthier, tastier, more expensive and better for the environment than conventional food. Also women were more positive about organic food than men. Similar to Lockie et al. (2002), these researcher reveal that unavailabilty of organic food is one of the major barriers of organic food consumption in Australia. Consistent with many other international studies, Chang and Zepeda (2005) suggest that organic food consumption among Australians is motivated by concerns regarding health and the environmental wellbeing.

A very recent study has revealed that the majority of Australian organic food consumers are females aged between 25-55 years (Oates et al. 2012). In this same research, and similar to many international research findings, Australians generally consume more organic fruit and vegetables as compared to organic meat products. Researchers in this field however point out the importance of further investigation into organic food consumer buyer behaviour. In particular, they suggest that a clearer identifiable profile of the organic food consumer in Australia should be agreed upon, which will undoubtedly resolve the ongoing debate relating to the health benefits of organic food.

## The certification of organic foods

To purchase organic food, it needs to be qualified as 'organic' and therefore needs to be certified by a duly constituted certification body or authority as noted in the previous definition. However, across many countries there exist inconsistencies in organic food labelling and certification systems, and this may adversely affect the purchase intentions of prospective consumers (Bonti-Ankomah and Yiridoe 2006). In this context it has been demonstrated that although 38.5 per cent of survey respondents were interested in sighting organic labels (e.g., EU organic label) when buying organic products, only 12.5 per cent of them were able to describe those labels (Pivato et al. 2008). Other researchers concluded that although many consumers emphasise the use of certified organic symbols in guiding organic food consumption, there is generally a lack of trust in those certification schemes (Bhaskaran et al. 2006). These researchers also highlight the existence of multiple certification schemes, which obviously erodes the confidence of consumers. Further, the use of diverse terminology such as organic, green and environmentally friendly whilst promoting organic food products, contributes to consumer mistrust (Zhao et al. 2007).

There are however other inhibitors (and motivators) impacting consumers' buyer behaviour of organic food. As noted earlier, there co-exists in the literature, evidence of favourable consumer attitudes toward certified organic food (e.g., Kihlberg and Risvik 2007; Aschemann et al. 2008) and also evidence of deterrents to the actual purchases of organic food (Sahota 2009; Schaack and Willer 2010). In particular, Fearne (2008) reports that repeat purchases of organic food is considerably lower than that for conventional food, and this hampers the growth of the organic food industry. More importantly, Fearne (2008) claims that consumers' expectations of organic food is not being met, especially when it relates to its taste. Besides the contentious issues relating to the certification of organic food as mentioned previously in this section, there are various other issues which contribute towards the gap between attitudes during pre-purchase evaluation and actual purchase behaviour of organic food.

As acknowledged by Loughnan (2012), Genetically Modified (GM) food is nothing new. It has been part of the added new engineering techniques adapted by scientists and farmers over the centuries. In Australia, GM crops have not become mainstream for the moment but some argue it's a matter of time. Some experts suggest that sooner or later GM agricultural production will see a much larger range of fruit and vegetables become GM processed. What many observers, including Loughnan, have sought to raise is that there is much need for further research on this matter especially when it comes to its impact on human health. Other chapters of this book will address some deeper aspects of GM farming and its consequences.

**What is in the book**

This book is a collection of essays from a range of specialists in diverse fields of organic food as well as areas which are not strictly organic food related. The intention of this book was to provide a more scholarly outlet for those that wish to broaden and inform their readers of the meaning and facets of the organic food industry. The chapters have both an international and an Australian breadth and we have allowed the authors to provide their research and studies which may straddle across both terrains. It is in our view only natural that case study examples and direct experience will be focused on the market we are most familiar with and that is the Australian one. Therefore some of the focus will be on Australia.

This study is not an advocacy of organic foods and related products though it does not shy away from an open discussion of this most controversial industry. The purpose of this book and most of the chapters inside it, is to shed light on what has been a 'shadowy' existence and openly examine its merits or otherwise. A number of chapters in this book openly dispute or take to task the role of organic food and the organic industry. On the chapter the 'Science of organic food' readers will be subject to a sharp contrast of views and in some cases even powerful evidence on the strength of the nutrition argument, along with other health and ecological considerations. Readers will find that there are counter posing views and even counter posing evidence to argue the two sides of organic food. They will find statements of the kind:

> Research has shown certified organic foods can be more nutritionally dense than their non-organic counterparts and deliver more essential nutrients per calories consumed. International studies have found certain organic fruits, vegetables and grains can contain significantly higher concentrations of health-promoting polyphenols and antioxidants – both of which have been linked to the prevention of cardiovascular disease, cancer and osteoporosis (BFA 2013).

At the same time other studies and authors have indicated an absence of evidence to justify assertions about the greater nutritious nature of organically farmed products. Some of this debate has been influenced by the role played by the large retail outlets which have been both interested in organically farmed products but also in retailing higher quality conventionally processed agricultural products. Some have highlighted the ability of these retailers on being able to play off farmers and wholesalers against each other to secure a higher quality crop or product. None of these improvements however come at zero cost. The expenses of achieving higher quality crops have been significant and affected purchasing habits of consumers.

Also within the study, readers will have the opportunity to become familiar with other wider ranging and related themes which will include: the history of organic food within Australia (John Paull), organic food and consumer behaviour and specifically the role of LOHAS (Antonio

Lobo), the supply chain of organics (Chandana Hewege) and an examination of the organic industry in Australia (Andy Monk). On the international side of this industry the reader will be able to examine the expertise of international specialists and their evaluation of the trends in the international organic market (Andre Leu) and a very important case study of China (Jue Chen and Barry O'Mahony). In the last section of this book the study addresses the very topical and central issue of food security (Mark Gibson) and the many dimensions that should be examined when looking at food security. The final chapter of this study looks at the political and social nature of food sovereignty (Alana Mann) and its renewed importance in creating a balance between the small agricultural producer and the large multinational food processors.

## Conclusion

The organic industry, of which organic food is the main component, is clearly highly political in character both because of the product, the players, the industry philosophy and to some extent the dispute over the 'science of organic food'. This chapter has tried to define organic food, examine the historical evolution of organic food production and chart the direction of organic food internationally as well as in Australia. This chapter has also charted the well cited definitional constructs of the characteristics of organic food which have made it unique and not surprisingly controversial.

The characteristics of organic farming are largely centred on zero chemical use in its production systems, which is also the general focus of many organic food certification schemes. The book also tackles the controversial and possibly inconclusive debate about the science behind organic food. Some countries, especially those that have partially rejected or raised concerns about modern industrial production are looking at organic products in ways they never did in the past. In addition some nations like China have embraced organic dairy as a response to a loss of trust of Chinese local production of the same as a result of scandals revealing toxic additives to these dairy products. New concepts of organic production are equally changing the face of organic food consumption and in some arenas the lifestyle driver is playing a significant role in promoting organic products. The emergence of debate on food security and its many derivatives as well as food sovereignty opens up the area of food, and not just organic food, to many important facets which are societal issues in need of rectification and addressing. While this book has concentrated much of its content in an Australian context we have also sought to draw many international experiences and case studies into this analysis where they are relevant.

As readers will see, matters related to food and organic food traverses many social, economic and political boundaries and often the boundaries are not clearly delineated. Nonetheless organics has captured the attention of food experts and policy makers and is in need of further more informed debate and discussion. For those who advocate organic food this study will aid the cause of understanding the benefits of this industry and providing a more scholarly discussion on the worth of organic food. For those who are sceptics of this industry this study we hope has made available other viewpoints and findings. Most importantly this study is designed to elevate the debate such that policy makers become sensitised to alternative claims made about this industry and not continue to shy away from the need for more clarity. This book is fully aware of the assertions made about the benefits of organics as well as the counter claims by those less supportive. This current book seeks, even if minimally, to contribute to a more informed understanding on the organic industry in a transparent and open manner. We hope it also leads to many others who will follow in our footsteps.

# 2
# The science of organic food

*Bruno Mascitelli*

## Introduction

The organic food sector is growing worldwide and doing so rapidly, such that it is transforming from a niche boutique status to an established sector with both the number of organic consumers and producers worldwide growing. It will be often stated that along with this increased consumer interest in organic food comes the belief that organic foods are better than conventional foods and are sometimes considered to be healthier, safer, and taste better than conventional foods (Loughnan 2012). On this more will be said later in this chapter.

Yet, like climate change, the notion that organic food is healthier is not a universally held view. While some consumers and organic practitioners require no convincing about the health benefits of organic food, a considerable number of other stakeholders take issue with these presumed assertions on the apparent health benefits. As there are numerous other benefits which are all ascribed to the use of organics, there is a clear need not only to engage in an open and necessary debate, but also to establish the scientific basis of organic food and organic products in general. It would be inaccurate to state that there is a raging debate on the scientific component of organic products. However, there are significant sectors of the food industry, of society and consumers with strongly held alternative views, and as such opposed to the 'scientific' reasoning about the health benefits of organic products.

As with the growth of many other industries, gaining the support of stakeholders (national governments, policy makers, consumers, farmers and other institutions) is essential in ensuring the growth of the organic food industry. The support of these stakeholders depends on the conviction that the organic food industry is beneficial and hence, essential. These arguments should be built on compelling evidence that consumption and production of organic food provide us with benefits such as environmental and human wellbeing or other benefits. As such, similar to other unresolved social discourses, providing scientific evidence as to how organic food could be healthier than conventional food is important.

Surprisingly the debate on organics and its benefits has largely been limited to one that is preaching to the converted. Equally those that are sceptics have on the whole debated their scepticism in limited ways and to limited audiences. While this study cannot pretend to be the ultimate forum for this engagement, it realises that the debate has largely been one like boxers sparring from a distance thereby limiting the debate. This chapter will not pretend that it can definitively resolve this debate and controversy but will at least raise some of these questions to a wider public and engagement.

Providing scientific evidence of the health benefits of organic food may increase and retain consumer demand for organic food. From a producer perspective, providing scientific evidence as to how organic agricultural practices ensure the environmental wellbeing (e.g., biodiversity) is essential in gaining the support for infrastructure development and regulating the organic food industry. As such, it is timely to have a discussion on the science of organic food.

Australian conventional agriculture uses hundreds of thousands of tonnes of synthetic or artificial fertiliser on crops each year. Despite this and the fears that many might have about food safety, Australia has a relatively safe record and in 2010 Australia was ranked the safest of the OECD countries (Loughnan 2012, p. 34). This chapter discusses the scientific evidence based justification of health or other benefits of organic food and agricultural practices. In alignment with this purpose, this chapter addresses three key areas. The first section discusses the science of organic food from a consumer perspective. The second section discusses the science of organic food from producer perspective followed by the chapter conclusion presented in the third section.

As noted in the literature review of this book, definitions of organic food largely focus on the mode of farming and agricultural practices. In particular, many definitions encompass not using chemical substances and fertilisers in organic food production. On this overall definition there seems to be some universal consensus. At the same time the definition of organics does not in any way address environmental and human wellbeing – but simply the absence of chemical substances. Further, health or medicinal benefits of consuming organic food and other scientific information on organic food consumption or agriculture are largely ignored or are not taken seriously. Many researchers and practitioners have been arguing on the aforementioned phenomena since the 1920s (Rosen 2010). Therefore the debate has a bi-dimensional nature – consumer perspective and producer perspective. The section that follows will provide for the reader the highlights of the debate on the science of organic food along these lines.

In late 2012 the USDA (United States Department of Agriculture) produced a dossier of 13 reasons why organic was better than conventional food (USDA 2012). It is reproduced here to allow readers to see the way in which a government ministry couched the debate and the scientific benefits of organics.

The authors of this dossier through the auspices of the USDA have presented a series of wide ranging series of reasons, though not necessarily a supremely convincing argument of the superiority of organic food. Some of the experts in the Health Science of diet have refused to give a blanket statement of endorsement of organic food

## Table 2.1: 13 reasons why organic is better than conventional food

*Making the commitment to organic food hardly means a sacrifice. The benefit to your personal health is just the start. Here are 13 reasons that take you far beyond the standard ones you most likely already have heard.*

1. Stringent Standards – The 'Organic' certification you see on a product means that it has been grown, processed and handled according to strict guidelines and procedures – at the highest level it means it contains no toxic chemicals. The federal government set standards for the production, processing and certification of organic food in the Organic Food Production Act of 1990 (OFPA).

2. Free of Genetic Modification – Organic food cannot be grown using genetically engineered seeds. Why would a seed be genetically altered? Our most common crops – corn, soy, cotton and canola – are often sprayed with heavy doses of pesticides that would otherwise even be damaging to the crops if their genetic structure had not been engineered to withstand these chemical substances.

3. Better for the Soil – Organic farming returns nourishment to the soil, which in turn creates better conditions for crops to thrive during droughts. Healthy soil acts much like a sponge and filter, so it helps to clean the water passing through it. Organically treated soil also traps carbon – and less of it in our atmosphere means fewer effects from climate change.

4. Better for the Water – Organic food is grown without chemical fertilizers or pesticides, which can be leached from the soil and end up in water supplies.

5. Innovative Research – Mostly at their own expense, organic produce growers have paved the way with innovative research that has created ways to reduce our dependence on pesticides and chemical fertilizers – both often by-products of fossil fuels. They also have led the way in developing more energy efficient ways of farming.

6. Increases Biodiversity – For decades now, organic farmers have been collecting and preserving seeds – as well as reintroducing rare or unusual varieties of fruits and vegetables. It is mostly thanks to organic farmers that a large selection of heirloom tomatoes has become a common occurrence in grocery stores.

7. Increases Consumer Choices – Thanks to organic farmers, nearly every food category offers an organic alternative. That has even extended into textiles. You can now sleep on sheets and wear clothes made of organic cotton.

8. Harmonious with Nature – The organic philosophy of growing believes that wildlife is integral to a farm. Organic farms do not displace wildlife – they embrace it.

9. Protects Family Farms and Rural Communities – As U.S. farm production continues to consolidate, small family farms are in danger of disappearing. With its higher profit margins, organic farming may be one way for family-owned farms to thrive – and in turn, revive dying rural communities.

10. Fresher, Better Tasting – There's no argument that fresh food tastes better. Organic food often is fresher because is more perishable and has a shorter shelf life.

11. Part of Your Community – Because organic food contains no preservatives and must be consumed quickly, it often is sold to grocery stores nearby. Chances are, if you eat an organic vegetable or fruit, you wouldn't have to travel far to visit where it was grown.

12. Sustainable Seafood Choices – Visiting organic markets allows you to purchase seafood that is still abundant and fished or farmed in environmentally friendly ways.

13. Safer and More Humane Animal Products – Organically raised animals are not fed animal by-products, or given antibiotics and growth hormones. And to increase their health, they're given more room to move as well as access to a natural outside environment. Crowded living conditions are a leading cause of animal sickness and suffering.

*Source: Blake 2012.*

as being necessarily more nutritious and healthier to consumers. In Loughnan's study she indicates:

> The Dieticians Association of Australia says the jury is still out on whether organically grown food is superior to conventionally grown food and that the topic needs far more research. Until conclusive evidence is found to prove or disprove the claims made for organic food, the Association says both organic and conventionally grown foods can provide all the nutrients required when included in a healthy balanced diet (Loughnan 2012, p. 31).

There is a long-standing debate on the health benefits of organic food. In addressing the notion that organic food had extra nutritional value, a review in 1997 of 150 publications found that there was no difference between the nutritional values of organically and conventionally grown vegetables (Woese et al. 1997). The findings of this study were later supported by other researchers and this study has resonated throughout the industry (Magkos et al. 2003; Kristensen et al. 2008). Brandt and Mølgaard (2001), in direct contrast with the above findings found that on average, organic vegetables and fruits contained more vitamins, minerals and carbohydrates than conventional vegetables. It is primarily due to this conflicting evidence about whether organic foods have better nutritional values than conventional foods have, that an open and honest debate is required within the scientific community as well as in the industry.

## The science of organic food from the consumption perspective

Our modern food supply is vulnerable and can be threatened for many reasons. Some of the sceptics of organic food point out that the supporters of organic food and organic living reason with a fundamental mistrust of science and technology. Many in the science area believe that NGOs, anti-GM organisations and those concerned with food miles, are able today to exert a significant impact of their views on the media and on the thinking of both consumers and policymakers. Scholars of science often complain that the scientific reality is inadequately presented to consumers. Gibney, a well-known scientist in the food area states candidly and recently renewed his long held view that "Organically grown food is nutritionally identical to conventional food" (Gibney 2012). Gibney when examining his own country's approach towards organics stated:

> The UK Food Standards Agency commissioned a review of all the literature relating to the nutritional quality of organic food. The review concluded: 'On the basis of a systematic review of studies of satisfactory quality, there is no evidence of a difference in nutrient quality between organic and conventionally produced foodstuffs' (Gibney 2012).

Many researchers agree that consumers' health and safety concerns have partially driven organic food consumption (e.g., Williams and Hammitt 2001; Sargeant et al. 2006). However, it is less certain if consumers have a complete understanding of scientifically proven benefits of organic foods. While demand for organic food has increased, some consumers do not clearly understand the exact difference between organic and conventional foods. As such, some researchers find that consumers often confuse organic and free-range products because they believe that 'organic' is equivalent to 'free-range' food (Harper and Makatouni 2002). Moreover the differentiation of organic from conventional food such as milk for consumers can be discerned, as Hill and Lynchehaun (2002) found, from the key difference in the taste and the cost. This only reaffirms the realisation that many consumers do not have a complete understanding of the benefits of organic foods and how organic foods can be differentiated from conventional foods.

Although systematic studies that synthesise the scientific evidence of organic food consumption are scarce in previous literature, some recently published studies have sparked off the long-standing debate on health benefits of organic food. A recent study claiming to be the world's biggest research project on organic foods, reviewed research articles on organic food consumption that were published in the last

50 years during 1958-2008 (Dangour et al. 2009, 2010). Dangour and others reviewed 52,471 articles in 2009 and 98,727 articles in 2010. The study was organised by the London School of Hygiene and Tropical Medicine and funded by the Food Standards Agency in the UK. It was found that except for some evidence that organic food consumption reduces risk of eczema in infants, there was no compelling evidence of nutritional benefits (e.g., vitamin C, calcium and iron) and health benefits of consuming organic foods. This research finding challenged to the core the argument that organic foods have more health benefits than conventional foods. The organic food industry in the UK was not impressed and the research findings were heavily criticised by many of organic food campaigners. One such campaigner, Peter Melchett, who was a policy director at the UK Soil Association criticised the research methodology adopted by Dangour's research group. He noted:

> We are disappointed in the conclusions the researchers have reached. It doesn't say organic food is not healthier, just that, according to the criteria they have adopted, there's no proof that it is (The Soil Association, 2009).

The Soil Association is a British lobby group in the organic industry and a certification body of organic food. This association had been playing a significant role in promoting organic food in the UK since 1980s. Along with the aforementioned criticism against Dangour's study, the soil association called for more research on health benefits of organic food. According to the Food Standards Agency, which funded Dangour's study, in their view the long-standing debate on whether organic foods have more health benefits than conventional foods is over.

According to Rosen (2010), organic food campaigners such as the Soil Association rely on studies that have neither been reviewed by independent scientists nor the findings of those studies have been statistically verified. More importantly, Rosen (2010) claimed that organic food campaigners tended to ignore the research findings that did not support their views. Based on this claim, one may easily reject the criticisms of those organic food campaigners and agree with Dangour's study findings. Taking a definitive stand in favour of these findings might be easy but possibly too early as not only organic food campaigners but also some scientists disagree with Dangour's findings (e.g., Benbrook et al. 2009). In an editorial piece Charles Benbrook (chief scientist at The Organic Centre of a not-for-profit research and education organisation in the USA) criticised Dangour's research group for excluding 54 per cent of studies that could have been considered for the analysis to be able to then reach an even better finding.

In 2008, Benbrook and other scientists found significantly different results from those of the Dangour's study (Benbrook et al. 2008). The researchers found that organic foods contained, on average, 25 per cent higher concentrations of nutrients than conventional foods. As such, it is apparent that the long-standing debate on the health benefits of organic good is still going on. Emerging new yet contradictory research findings on organic foods also seem to reignite the aforementioned debate. For example, a very recent study systematically integrates and compares the finding of three previous studies that focused on the nutritional benefits of organic and conventional dairy products (Palupi et al. 2012). This meta study shows that organic dairy products contain significantly higher protein and other nutritional benefits (e.g., omega-3 fatty acid, *cis*-9, *trans*-11 conjugated linoleic acid, *trans*-11 vaccenic acid, eicosapentanoic acid, and docosapentanoic acid) than those of conventional dairy products. Similarly, it is also found that organic vegetables have more micronutrient contents than conventional vegetables (Hunter et al. 2011).

In 2011, a review study was conducted with a purpose of postulating a framework for estimating the scientific impacts of previous research findings that focus on health benefits of organic food (Huber et al. 2011). The researchers also addressed the issues of diverse methodological approaches adopted in previous research. According to this study; (a) a considerable number of previous studies found that organic foods have lower nitrate contents and less pesticide residues than conventional foods and (b) organic foods usually have higher levels of vitamin C and phenolic compounds. The researchers, however, claim that there is a very high variation among the findings of comparative studies. Moreover, the researchers found that there is no relationship between nutritional value and health benefits. The difficulty of drawing conclusions from analytical data about the health benefits of organic foods is also shown.

According to some other researchers, drawing clear conclusions on the health benefits of consuming organically produced foods are difficult due to (a) the lack of a clear operational definition of health and (b) the inability to distinguish between different levels of health using valid biomarkers (indicators of a biological state), (Huber et al. 2012). Adding further evidence to the difficulty of making conclusive comments on the health benefits of organic food, a recent study also found mixed results. Based on a survey with 566 respondents the study found that whilst 30 per cent of the respondents reported no health effects, the majority of the other respondents (70 per cent) reported better general health, including feeling more energetic and having better resistance to illnesses. Further, a positive effect on mental wellbeing (30 per cent), improved stomach and bowel function (24 per cent), improved condition of skin, hair and/or nails (19 per cent), fewer allergic complaints (14 per cent) and improved satiety (14 per cent)(van de Vijver and van Vliet 2012).

Another study analyses the differences in mineral composition between conventional potatoes and organic potatoes (Griffiths et al. 2011). The researchers found that although organic potatoes had more copper and magnesium, less iron and sodium, they had the same concentration of calcium, potassium and zinc as conventional potatoes. As such, the researchers concluded that although these differences could minimally affect total dietary intake of these minerals, this intake was highly unlikely to show considerable health benefits. Supporting this claim, a recent study which investigated 63,561 pregnant women during the years 2002-2007 found that unlike other socio-economic characteristics (education and income) of pregnant women, frequent organic consumption cannot be directly related to health benefits (Torjusen et al. 2010). It was therefore suggested that personal and socio-economic characteristics are important factors and need to be included in future studies of potential health outcomes related to organic food consumption during pregnancy.

In summary, providing scientific evidence of health or other benefits of organic food is essential in encouraging organic food consumption and providing evidence that organic food has proven benefit over conventional food. But as has been shown there are inconsistencies and some of the controversial research findings on the health benefits of organic food which keep the question wide open. There continues, based on research findings, difficulty of finding a close relationship between organic food consumption and health benefits.

## The science of organic food from the production perspective

This section describes scientific evidence based debates on the benefits of organic farming or agricultural practices. A few special remarks should be added at the beginning of this section. As noted in the previous chapters of this book, many widely cited definitions of organic food are centred around organic farming or agricultural practices. For example, organic agriculture is defined as a farming system that maintains and enhances the health of soils, plants, animals and humans (IFOAM 2006). IFOAM objectives and regulations seem to formulate the definitions of organic food that are also centred on quality of organic food in terms of process and product related aspects (e.g., Kahl et al. 2011). As also noted in the previous chapter of this book, organic agriculture is viewed as an alternative farming system as well as a social movement against agri-industrialisation. This section especially focuses on the debate on scientifically proven benefits of organic farming and agricultural practices.

Studies that investigate scientific evidence of organic food and agricultural practices are often organised as comparative studies of organic and conventional agricultural practices. The studies that focus on soil quality (e.g., Mäder et al. 2002; Fliessbach et al. 2007; Grandy and Robertson 2007), biodiversity (e.g., Bengtsson et al. 2005; Hole et al. 2005; Tscharntke et al. 2005) and green gas emissions (e.g., Flessa et al. 2002) largely compare the environmental impacts of organic and conventional agricultural practices.

A meta review study of literature published between 2002-2005 finds that organic farming increases species richness. In particular, it is found that organic farming practices ensure on average 30 per cent higher species richness than conventional farming practices (Bengtsson et al. 2005). The researchers, however, claim that this finding varies among studies. They also suggest that although positive effects of organic farming on species richness can be expected in intensively managed agricultural landscapes, such benefits cannot be expected in small-scale landscapes.

## Pesticides

For a product to be classified as organic there should be no synthetic residues on the product or applied to it. In contrast, conventionally grown foods may contain trace levels of pesticides. The Department of Agriculture of most countries conduct surveys of crops and analysis of samples for residues of pesticides as a matter of course. It is normal to find that crops which have been applied with pesticides of levels between 0.1 per cent and 10 per cent of the safe exposure dose established by the World Health Organisation are considered to have zero public health risk from a pesticide use. As to be expected this finding is not accepted by all.

Kouba (2003) found that organic farming itself does not necessarily guarantee the absence of contamination from pesticides, mycotoxins, bacteria, and parasites. This study also went on to affirm that organic farming on the other hand ensures environmental wellbeing. Another review study compares the impacts of organic agricultural practices with that of conventional agricultural practices on biodiversity (Hole et al. 2005). This study finds that organic agricultural practices generally ensure greater biodiversity than conventional agricultural practices. More importantly, the study highlights several key issues to be taken into consideration in making a conclusive comment on the benefits of organic agricultural practices. Firstly, the researchers highlight the uncertainty about whether organic agricultural practices would ensure greater biodiversity than conventional agricultural practices would especially when the latter involves small areas of cropped habitats. Secondly, the researchers show that methodological issues in the existing comparative studies of organic and conventional agricultural practices negatively influence building a strong argument for either practice. Thirdly, they also stress the lack of understanding on the impacts of organic farming in pastoral and upland agricultural practices and hence, call for longitudinal studies to address the aforementioned issues in comparative studies.

Using systematic review and meta-analysis methodology, a recent study compared the organic and conventional poultry, swine and beef production (Young et al. 2009). The specific purpose of the study was to compare the prevalence of zoonotic (a bad bacteria that could be transmitted between species) and potentially zoonotic bacteria resistant to antimicrobials (an inhibitor of the growth of disease-causing microorganisms) in organic and conventional poultry, swine and beef production. The study found no difference in the prevalence of Campylobacter in organic and conventional retail chicken. Further, the study highlighted the limited and inconsistent research findings on the prevalence of bacterial enteropathogens and potentially zoonotic bacteria in other food-animal species (Young et al. 2009).

Rundlöf and Smith (2006) address another question on whether organic agricultural practices actually preserve biodiversity. They investigated the effects of organic agricultural practices in different landscapes (homogeneous and heterogeneous landscape diversity) on butterfly species. They found that both organic agricultural practices and landscape heterogeneity significantly increased butterfly species richness. In addition, the study found that there was a significant interaction between organic agricultural practices and landscape heterogeneity. Put simply, the positive effect of organic agricultural practices on butterfly species richness could only be ensured in homogeneous landscapes.

Bellon and Lamine (2009) carried out further studies in which they combined agricultural and social scientists' viewpoints on organic and conventional farming practices. These researchers focused on cause and effects of transforming from conventional agricultural practices to organic agricultural practices. They (a) highlighted the importance of investigating the transformation process as a multidimensional issue that relates to both production and social practices, and (b) identified organic agricultural practice as an approach that calls for macro changes in conventional agricultural practice ensuring environmental and social sustainability.

Another study examined on this abundant literature was that of comparing the environmental impact of organic and conventional farming practices (Mondelaers et al. 2009). The environmental impact under examination included land use efficiency, organic matter content in the soil, nitrate and phosphate leaching to the water system, greenhouse gas emissions and biodiversity. The findings of this study concluded that the soils in organic farming had on average a higher content of organic matter. It was also found that organic farming contributed positively to agro-biodiversity (breeds used by the farmers) and natural biodiversity (wild life). The study, however, concluded that there was no clear evidence in relation to the impact of the organic farming system on nitrate and phosphorous leaching and greenhouse gas emissions.

Another study on strawberries evaluating the three varieties of strawberries for mineral elements, shelf life, phytochemical composition, and organoleptic properties (Reganold et al. 2010) produced significant findings. The study analysed traditional soil properties and soil DNA. They found that although the organic farms had strawberries with longer shelf life, greater dry matter, and higher antioxidant activity and concentrations of ascorbic acid and phenolic compounds, those strawberries had a lower concentrations of phosphorus and potassium (Reganold et al. 2010). Similar to many other previous studies, this study confirmed that organic strawberries were sweeter and had a better flavour, overall acceptance, and appearance than their counterpart conventional strawberries. In regards to organic farming, the researchers found that the organically farmed soils had more total carbon and nitrogen, greater microbial biomass and activity, and higher concentrations of micronutrients. Organically farmed soils also exhibited greater numbers of endemic genes and greater functional gene abundance and diversity for several biogeochemical processes, such as nitrogen fixation and pesticide degradation (Reganold et al. 2010).

With regard to taste as a criterion of differentiation, the most comprehensive study carried out was at the University of Kansas, where experimental techniques allowed the researchers to vary the fertiliser type whilst growing the crops in identical microclimates. Trained taste panels could not distinguish the differences between organic and conventionally grown foods (Gibney 2012).

## Environmentally friendlier?

Organic agriculture is commonly believed to be more environmentally friendly than conventional agriculture. A report commissioned by the British Department of the Environment, in contrast to this commonly held view, concluded that was insufficient evidence available to state that organic agriculture overall would have less environmental impact than conventional agriculture. In support of this finding the report stated:

> In particular, from the data we have identified that organic agriculture poses its own environmental problems in the production of certain foods, either in terms of nutrient release to water or in terms of climate-change burden. There is no clear-cut answer to

the question: which 'trolley' has the lowest environmental impact: the organic one or the conventional one (Gibney 2012).

Another British finding on the same theme of the environment found that organic milk, cereals, and pork all generated higher greenhouse gas emissions than their conventionally farmed counterparts. Organic beef and olives produced lower emissions and that therefore organic was not demonstrably better for the environment (Hope 2102). For example, organic wheat production required less energy than conventionally farmed wheat but required more land. On the other hand there was no difference between organically grown or conventionally grown potatoes and both required 40 per cent of the total energy for storage.

## Healthier and tastier?

Lairon (2011) finds a positive relationship between organic agricultural practices and high quality standards of food. However, not all researchers agree with this finding. A study investigates the impact of organic and conventional agricultural practices on the content of carotenoids in carrots in human diets (Søltoft et al. 2011). It is found that the content of carotenoids in carrot roots in human diets has no relationship with the agricultural practice. According to the findings, although the plasma status of carotenoids increased significantly after taking organic and conventional diets, no systematic differences between the agricultural production systems are observed. Similar to aforementioned findings, another more recent experimental study investigates differences in nutritional values and health benefits of carrots from four different cultivation systems grown in two consecutive years (Jensen et al. 2012). In this field experiment, rats were fed a diet with 40 per cent carrot content. The study found no clear influence of cultivation methods or harvest year on the nutritional quality of carrots or effect of cultivation methods on health benefits (Jensen et al. 2012). Hence it can be seen that more recent research has found no significant differences in organic or conventional products, in relation to nutritional values or health benefits.

Even if organic isn't healthier, it is still tastier in some people's eyes. Researchers found no consistent differences in the vitamin content of various foods. They also found no difference in protein or fat content between organic and conventional milk (Hope 2012). They were also unable to identify specific fruits and vegetables for which organic appeared consistently to be the healthier choice. Researchers at Stanford University's Medicine school tackled the question front on. They concluded:

> Some believe that organic food is always healthier and more nutritious. We were a little surprised that we didn't find that (Hope 2012).

In the same research it was discovered that organic produce was 30 per cent less likely to be contaminated with pesticides than conventional fruit and vegetables, but not guaranteed to be totally pesticide-free, while pesticide levels of all foods came within the allowable safety limits (Hope 2012). While there is the strong perception that organic foods are going to be much better for us, and for our health, studies have found little evidence of a major difference. "The biggest health problems facing Australians are to do with over consumption of food, not inadequate consumption of fruit and vegetables" (Palmer 2012).

## Vested interests or a lost opportunity: An unhealthy debate over Genetically Modified (GM) food?

While recent headlines have made much of the rise of GM crops mutation producing the greatest fear and mistrust of the food industry, modification of crops began much earlier. As acknowledged by Loughnan, Genetically Modified (GM) food is nothing new (2012). It has been part of the added new engineering techniques adapted by scientists and farmers over the centuries. Conventional plant breeding preceded this in ways that are less known and no less relevant to this discussion. In 1946 Herman Muller was awarded recognition for the discovery that X-rays could induce genetic mutations in crops and in insects, thus dramatically speeding up the task of breeding. His achievement was explained as:

> Greatly simplified, X-ray irradiation, as also ionizing irradiation, could be likened in general to a shower of infinitely small but highly explosive grenades, which explode at different spots within the irradiated organism. The explosion itself tears the structure of the cell to pieces or disturbs its arrangement. If such an explosion happens to take place in or close to a gene, its structure, and therewith also its effect on the organism, may be changed (Muller 1946 cited in Gibney 2012).

Under the auspices of the International Atomic Energy Agency (IAEA) in Vienna, the use of atomic radiation to induce mutations is a completely normal activity in plant breeding. The genetic changes lead to mostly useless and sterile mutants, but when you expose thousands of plants to this radiation a handful will show some desirable trait. This technology has given us basmati rice, red grapefruits, barley for the whiskey industry and many thousands of crops, all listed on the IAEA website. Modern genetic engineering alters the plant genome with great precision compared to the approach of radiation breeding where thousands more genes are altered. While GM foods have produced great concern and fear little concern was raised about the radiation-bred crops or harm to biodiversity or that it would cause genetic drift, or other horrors of modern GM. The US National Academy of Science undertook a risk assessment of GM and radiation breeding and concluded that:

> Based on a detailed evaluation of the intended and unintended traits produced by the two approaches to crop improvement, the committee finds that the transgenic process presents no new categories of risk compared to conventional methods of crop improvement but that specific traits introduced by both approaches can pose unique risks (Gibney 2012).

Many European countries and in particular the UK, have refused to embrace GM crops and equally consumers all over the world have shown some reticence in this technology also. On the other hand the US and a number of Asian countries have placed greater trust in GM production and approximately 65 per cent of foods in US supermarkets have some form of modified ingredients (Loughnan 2012, p. 97).

In Australia, as can be evidenced from Table 2.2, GM crops in Australia are few and far between. On the whole the introduction of GM crops in Australia has been limited to canola (for cooking) and some cotton. On the other hand Australia does permit the sale and use of GM products grown in other countries. Some experts have posited that it is only a matter of time before the GM agricultural production will be open to many more products as a range of pilot arrangements in fruit and vegetables proceeds apace. Recently the *Financial Times* reported on GM crops stating "Like it or not, genetically modified crops are taking over the world" (Lex Column 2013). Interestingly the developing economies are playing a major role in allocating land for GM use. The strong position of Brazil, Argentina, India and to a lesser extent China

**Table 2.2: Genetically modified crops (million hectares)**

| Nation | 2009 | 2010 |
|---|---|---|
| US | 64 | 68 |
| Brazil | 22 | 25.05 |
| Atgentina | 22 | 23.50 |
| India | 8.5 | 9.5 |
| Canada | 8.4 | 9.1 |
| China | 4.0 | 3.8 |
| Paraguay | 2.7 | 2.8 |
| Pakistan | 0 | 2.7 |
| South Africa | 2.6 | 2.6 |
| Uruguay | 1.1 | 1.9 |
| Bolivia | 1.0 | 1.3 |
| Australia | 0.2 | 0.6 |

*Source: Lex Column 2013*

seem to be testimony to this as evidenced from Table 2.2. The main companies in this GM production include Monsanto, DuPont and Dow – all US multinationals.

The ethical debate about GM foods of recent is shifting away from just the debate about the 'Frankencorn' product but to the effectiveness of GM products in dynamically evolving and changing agricultural conditions.

## Conclusion

The science of organic food is, as can be evidenced from this chapter, an unresolved and contentious matter. Organic food, slow food, local food and natural food are choices people make but their choices should never engender a fear among ordinary consumers that conventionally grown foods are in any way inferior. The long-standing debate on the health benefits of consuming organic food are centred on claims and counter claims about the health benefits of organic food through the scientific evidence, or lack thereof. Moreover some of this debate centres on the difference between organic and conventional farming practices and the effects of additives, preservatives as well as other chemical additions to the state of the agricultural product.

While this chapter has provided numerous studies on one side or the other it is also clear that this long-standing debate continues to remain an open question until there is sufficient and convincing studies across a breadth of aspects of organic products before the health benefits of organic food are accepted. Methodological issues in the existing research on organic food and agricultural practices were also highlighted and to some extent remain incomplete. As noted at the beginning of this chapter, investigating scientifically proven benefits of organic food and organic agricultural practices is essential in gaining the support from the stakeholders for developing the organic food industry. It might be as some have argued that the real issues with the modern food supply are nutritional in nature. The contrast between the growing numbers of obese people in economically developed societies and at the same time millions of people going hungry every day in less developed societies should surely be the main focus of this debate.

# 3
# A history of the organic agriculture[1] movement in Australia

*John Paull*

## Introduction

History is of necessity incomplete – and this history of organic agriculture in Australia is no exception. Nevertheless, this account begins in the 1920s which is six decades before other accounts which have been styled as the 'History' of 'Organic Farming in Australia' and which begin their accounts 'in the early 1980s' (e.g., Wynen, 2007, p. 119), and it starts two decades before other researched accounts which begin in the 1940s (e.g., Jones 2010; Paull 2008, 2009).

Australia has some 'bragging rights' in the world of organic agriculture. The area of certified organic agriculture in Australia is 12,001,724 hectares which far exceeds that of any other country and accounts for 32 per cent of the world total (Willer, Lernoud and Kilcher, 2013). Australia was the home of the first 'organic farming' association and produced the first 'organic' farming periodical by an organic association, and produced the first statement of organic farming principles (Paull 2008).

Nevertheless, as a general rule, the organic agriculture movement in Australia has been a 'fast follower' of ideas that originated elsewhere and

---

1 Thank you for the access to manuscript material and records held at the Archives of the Goetheanum (Archiv am Goetheanum), Dornach, Switzerland, the archives of the Soil Association, Bristol, UK, the Archives Office of Tasmania, Hobart, Australia, and the archives of the Soil Association of South Australia, Adelaide, Australia, and thank you to the Institute of Social and Cultural Anthropology, the University of Oxford, for facilitating this research.

that have rapidly diffused internationally. The present account covers ten decades of the development of organics in Australia and traces the movement from its early infrastructure of resistance to the later infrastructure of capitalism – from meetings, newsletters, festivals and proselytisation, through to standards, labelling, certification, monetisation, and corporatisation.

Australia's involvement in the organic movement can conveniently be considered as four 'waves' of activity. In the present account, the First Wave (1920s and 1930s) is anchored by Rudolf Steiner's 1924 call at Koberwitz (now Kobierzyce, Poland) for a differentiated agriculture. This account reveals that Australian anthroposophists responded to this call by joining Steiner's Agricultural Experimental Circle (AEC) which was coordinated from Dornach, Switzerland. This First Wave culminated with the 'coming out' of biodynamic agriculture in 1938 – internationally with the publication of Ehrenfried Pfeiffer's book Bio-Dynamic Farming and Gardening, and in Australia with Bob Williams presenting the first public lecture on biodynamics at the home of Walter Burley and Marion Mahoney Griffin.

The Second Wave of organic agriculture in Australia (1940s and 1950s) is anchored by the coining of the term 'organic farming' in 1940, in England. This Second Wave witnessed the founding of the first associations in Australia dedicated to the advocacy of organics. It begins with the Australian Organic Farming and Gardening Society (AOFGS) founded in 1944 in Sydney, and it culminates with the year-long tour of Australia in 1959 by Eve Balfour, the founder of the UK's Soil Association.

The Third Wave (1960s and 1970s) is anchored by the publication of Rachel Carson's book *Silent Spring* in 1962 which breathed new life into the organics movement worldwide. A plethora of new associations and periodicals for the promotion, advocacy and exploration of organics appeared in the two decades that followed Carson. There was the publication of Australia's first popular and widely distributed book devoted to organics, and there were fledgling moves to develop organic standards, labelling and certification.

**Figure 3.1: World map of organic agriculture with countries proportioned according to the tally of certified organic agriculture hectares**

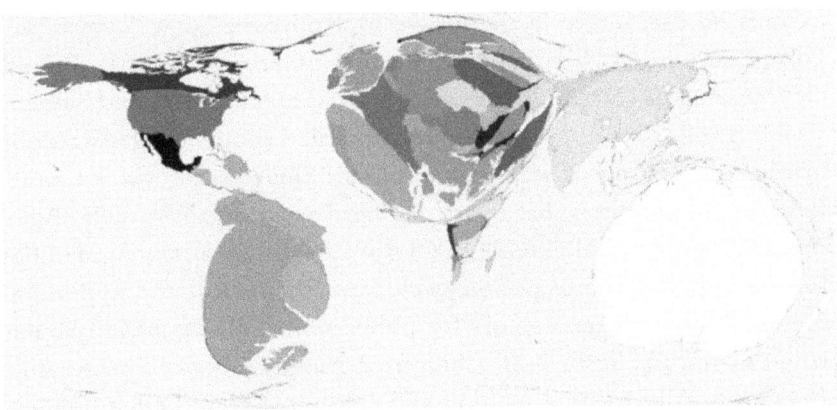

*Source: Paull and Hennig, 2013.*

The Fourth Wave (1980s to present) is anchored by the Chernobyl nuclear accident in Ukraine on 26 April 1986. Radioactive fallout spread across large swathes of Europe, and beyond, and this dramatically refocused the world's attention on the safety of its food supply. This Fourth Wave of the organics movement witnesses the maturing of organics thinking in Australia and the development of the apparatus of organics governance. The first organics certifiers were established along with the establishment of standards, logos, labeling and product differentiation. The International Federation of Organic Agriculture Movements (IFOAM) brought the fifteenth Organic Congress to Adelaide in 2005, Australia's first academic journal devoted to organic agriculture was established in 2006, and a national organic standard was implemented in 2009. In this Fourth Wave organics advocacy has become monetised and corporatised and Australia now leads the world with its tally of certified organic agricultural hectares (Figure 3.1).

## First Wave – Biodynamic pioneers – Anthroposophists – 1920s and 1930s

The Austrian philosopher and mystic, Rudolf Steiner, presented eight lectures on agriculture in the village of Koberwitz (now Kobierzyce, Poland) in June 1924 to 111 delegates from Germany (N=61), Poland (N=30), Austria (N=9), Switzerland (N=7), France (N=2), and Sweden (N=2) (Paull, 2011a). Farmers were concerned about changes wrought by the proliferation of synthetic fertilisers. Steiner's call was for a differentiated agriculture that treats the 'farm as an organism'. The Agricultural Experimental Circle (AEC) was established at the Agriculture Course at Koberwitz to put Steiner's 'hints' to the test and to develop the ideas into a form suitable for publication. This injunction culminated in the publication of Ehrenfried Pfeiffer's book Bio-Dynamic Farming and Gardening in 1938 in English, German, Dutch, French, and Italian (Paull 2011e).

In the interwar period about 800 individuals from around the world joined the Agricultural Experimental Circle (AEC) of Anthroposophical Farmers and Gardeners of the General Anthroposophical Society, which reported centrally to the Natural Science Section of the Goetheanum, Dornach, Switzerland. There were at least eleven Australian members of the AEC (Table 3.1).

The first Australian to join the AEC was Ernesto Genoni of Dalmore, Victoria, 70 km south east of Melbourne, about half way between Melbourne and Phillip Island. When Genoni enlisted in the Australian Imperial Force (AIF) on 25 January 1916 at Blackboy Hill, Western Australia, he declared that he was born in Tirano, Italy in October 1885, and he gave his age as 30 years, his nationality as Italian, his religion as Theosophy, and his next of kin as his brother Emil Genoni of Broome Hill, Western Australia. As a private in the 16th Infantry battalion of the Australian Army he embarked from Fremantle on 17 April 1916. He was discharged from the AIF "in consequence of joining the Italian

Army 18/10/16" (AIF, 1916). He joined the Anthroposophy Society in Milan, Italy, during Steiner's lifetime, on 27 September 1920, with his membership number as #7768. He returned to Australia after the war, and the Goetheanum Secretariat at the Goetheanum recorded his new address initially as 'Farm Dalmore, Gippsland, Victoria' and then later addresses at Dandenong (Vic), Katanning (WA), and Noble Park (Vic). Genoni joined the AEC receiving #165 of the Agriculture Course, in German, about twelve months before the first English translation arrived in Australia. Genoni's German edition was a numbered printed publication whereas the English edition of the Agriculture Course was issued in typescript up until 1938.

Table 3.1: Australian members of the Agricultural Experimental Circle of Anthroposophical Farmers and Gardeners (sorted in chronological order of sign on).

| Number | Member | Date | Address | Experiment location |
|---|---|---|---|---|
| 165 | Ernesto Genoni | c. Aug 1928 | Farm Dalmore, Gippsland, Vic | n/a |
| E14 | Crawford McDowell | 23/7/1929 | Sydney, NSW | Sydney, NSW |
| E17 | (signed by Ernesto Genoni for) Emilio Genoni | 15/10/1930 | Broome Hill, WA | Etna', Broome Hill, WA |
| E24 | Charles Burford | 7/5/1931 | Sydney, NSW | 'near Sydney, NSW' |
| E32 | Kenneth Milroy Temple | 12/11/1931 | Willoughby (Sydney), NSW | Sydney, NSW |
| E50 | Ruby A Macpherson | 26/3/1935 | Mt Tenandra, Gulargambone, NSW | Mt Tenandra, Gulargambone, NSW |
| E55 | Ruth Beale | 11/1/1936 | Annegrove, Mount Colah via Hornsby (Sydney), NSW | Mount Colah. NSW |
| E52 | Ileen Macpherson | 26/1/1936 | Dandenong, Vic | Dandenong, Vic |
| E66 | Robert Williams | 1/7/1937 | Willoughby (Sydney), NSW | Willoughby (Sydney), NSW |

| Number* | Member | Date | Address | Experiment location |
|---|---|---|---|---|
| E3 | Lucy C Badham | 7/11/1938 | Gordon (Sydney) NSW | |
| E4 | Eric M Nicholls | 9/2/1939 | ydney, NSW | Castlecrag (Sydney), NSW |
| E25 | Frank de Vere Kelly | 7/4/1939 | c/- Robert Williams, Willoughby, NSW | |

*Typescript copies of the Agriculture Course were numbered, they were issued in approximately numerical order, however copies returned to the Natural Science Section of the Goetheanum were occasionally reissued, hence the scrambled sequence reported here.

*Sources: original manuscripts signed by the members and held in the Archives of the Goetheanum, Dornach.*

Crawford McDowell of Sydney was the first Australian anthroposophist to receive the Agriculture Course in English. He joined the AEC on 23 July 1929, and the AEC membership was boosted to a dozen over the course of the following decade. The twelve AEC members were from three states: NSW (N=9), Victoria (N=2), and Western Australia (N=1). They each acknowledged receipt of a numbered copy of the Agriculture Course: "I accept it on loan for my own personal use in carrying out the experiments undertaken by <...name...> within the Agricultural Experimental Circle of the Anthroposophical Society, at the experimental station at <...place...>". They each undertook to maintain confidentiality: "I hereby undertake to preserve the strictest secrecy in all quarters as to the content of the aforesaid Lecture course. I will conduct the experiments in such a way as to exclude all possibility of imitation; and I undertake to lay the same obligations of silence on any of my fellow workers" (Paull 2011e, p. 25).

The Agricultural Experimental Circle was active in Australia for more than a decade (1928 -1939) with most members clustered in the northern suburbs of Sydney (in Castlecrag, Willoughby, Gordon and Mount Colah). At least five of the Australian AEC members, Ruth Beale, Lucy Badham, Robert (Bob) Williams, Eric Nicholls (architect and business partner of Walter Burley Griffin) and Frank Kelly (apiarist), were active members of the Castlecrag community of anthroposophists (Spathopoulos, 2007).

Biodynamic agriculture had its 'coming out' in 1938 with the publication of Ehrenfried Pfeiffer's book Bio-Dynamic Farming and Gardening which was released internationally and appeared in at least five languages: English, German, Dutch, French, and Italian (Paull 2011c). Steiner had said from the outset that this agriculture was for all farmers, but he imposed the injunction to put his 'hints' to the test and to then present this agriculture into a form suitable for publication (Steiner 1924b, 1924c).

Coinciding with the release of Pfeiffer's book, Bob Williams presented the first public presentation of biodynamics in Australia, on 26 June 1938, at the house of Walter Burley and Marion Mahoney Griffin at Castlecrag (Williams c. 1984). This public lecture marks the beginning of public advocacy in Australia for an agriculture differentiated from chemical agriculture.

The Griffins were enthusiastic anthroposophists. Marion Griffin provided the venue for Williams' lecture and later stated that: "Australia is rapidly being awakened to the fact that there is something radically wrong with the present methods. Scarcely a week passes without a column or half-column article about the serious condition of the soil, about blights of the products and the diseases of the animals fed from the large quantity production methods attained by chemical fertilizers and other materialistic scientific methods some of which are denuding the districts, some creating growing deserts whose dry sands are sweeping on, constantly increasing the desert area, some of which are reducing the fertility of the soil till it is becoming a pasty concrete-like substance, its fertility lost. It is a pity" (Griffin 1949, III p. 337).

In the 1930s, Anthroposophy Festivals were regular events at Castlecrag occurring four times a year: 'Easter, St John's Tide, Michaelmas and Christmas' (Spathopoulos, 2007, p.320). Festivals comprised plays (in German and English), eurhythmy performances and lectures. Marion

Griffin 'spoke on many topics' (Spathopoulos 2007, p.320). Recalling those times, Spathopoulos recalls:

> I remember young Bob Williams talking there, at one of the festivals, on organic farming and the wisdom of nature. He spoke of composting and the soil, of the phases of the moon and seasonal rhythms, and of interactions between the cosmos and the earth. He and his wife [Louise] later bought a couple of acres near the western boundary of Castle Cove, where they set up a small-scale experimental farm using the Steiner methods ... During the Second World War, Mrs Williams worked full-time on the farm as she raised their three children, while her husband returned to work there at weekends. They studied botany, especially grasses and medicinal herbs, and established a laboratory for developing biodynamic preparations for distribution in Australia, New Zealand and South Africa. Bob was sometimes invited to advise farmers on the best pastures to grow and gave lectures on related anthroposophical subjects (Spathopoulos 2007, p. 324).

These early anthroposophists set a ball rolling that rolls to this day. The biodynamic way of doing organics has maintained its identity and distinctiveness and has been incorporated into each of the organics standards up to and including the Australian Standard for Organic and Biodynamic Products, 6000-2009 (SA, 2009) (Table 3.2).

## Second Wave – Organics pioneers – 1940s and 1950s

The Second Wave of organics in Australia is predicated on a sequence of developments in Europe (Table 3.2). Lord Northbourne, a farmer in Kent and an Oxford University agriculturalist, was impressed with Ehrenfried Pfeiffer's knowledge. Pfeiffer's book Bio-Dynamic Farming and Gardening had appeared in 1938. Northbourne travelled to Switzerland to invite him to present Britain's first biodynamic conference. Pfeiffer accepted the invitation with the result that the Betteshanger Summer School and Conference on Biodynamic Farming was held at Northbourne's estate in Kent in the first week of July 1939. One of Pfeiffer's

## Table 3.2: Milestones of the development of organic agriculture in Australia (augmented with some contextual international events).

### First Wave – Anthroposophists – 1920s and 1930s

| Date | Event |
|---|---|
| 7-16 June 1924 | Rudolf Steiner (1861-1925) delivers the Agriculture Course at Koberwitz (now Kobierzyce, Poland) and founds the Agricultural Experimental Circle (AEC). |
| 28 May 1926 | Steiner's Agriculture Course published (in German) for members of the Agricultural Experimental Circle. |
| August 1928 | Ernesto Genoni (1885-1975) first Australian to join the Agricultural Experimental Circle of Anthroposophic Farmers and Gardeners. |
| August 1928 | Rudolf Steiner's Agriculture Course is released in typescript in English translation and made available to members of the Agricultural Experimental Circle. |
| 1929-1939 | Eleven more Australian anthroposophists join the Agricultural Experimental Circle and receive the English translation of the Agriculture Course. |
| 1938 | Ehrenfried Pfeiffer (1899-1961) publishes Bio-Dynamic Farming and Gardening, with editions in English, German, French, Italian and Dutch. |
| 26 June 1938 | Robert (Bob) Williams (1907-1984) presented the first public lecture in Australia on biodynamics at the house of Walter Burley and Marion Mahoney Griffin at Castlecrag (Sydney). |

### Second Wave – Organics Pioneers – 1940s and 1950s

| Date | Event |
|---|---|
| May 1940 | Lord Northbourne (1896-1982) published Look to the Land and coined the term 'organic farming'. |
| 4 October 1944 | Australian Organic Farming and Gardening Society (AOFGS) was founded. The association was based in Sydney, New South Wales (1944-1955). |
| April 1946 | First issue of Organic Farming Digest published by AOFGS 29 issues through to December 1954. |
| October 1945 | Victorian Compost Society founded (1945 - ≥1965). |
| 30 August 1946 | The Living Soil Association of Tasmania (LSAT) founded (1946-1960). |
| March 1952 | AOFGS publishes 10 principles of organic agriculture. |
| 5 July 1956 | Organic Farming and Gardening Society (Aust.) founded. |
| 1/11/1958-1/11/1959 | Eve Balfour (1899-1990), founder of the UK Soil Association, trip to Australia advocating organics. |

## Third Wave – Disseminators – 1960s and 1970s

| Date | Event |
|---|---|
| 1962 | Rachel Carson's book Silent Spring published |
| 1965 | Soil Association (South Australia Branch), later Soil Association of South Australia (SASA) (1975-2009), now OFA (SA) (Sept. 2009) |
| 14 December 1967 | Bio-Dynamic Research Institute registered (Alex Podolinsky, Powelltown, Vic.) |
| 1970 | Henry Doubleday Research Association of Australia Inc. (Sydney) founded |
| 1972 | Organic Gardening and Farming Society of Tasmania Inc. (OGFST) (1972-2009) |
| ≤ October 1972 | Organic Food Movement founded in Adelaide, SA (incorporated 9/6/1975) |
| 5 November 1972 | nternational Federation of Organic Agriculture Movements (IFOAM) founded |
| 1975 | Organic Gardening by Audrey Windram, first commercially published book in Australia dedicated to organics; royalties to the Organic Food Movement. |
| 1976 | Organic Farmer and Gardener periodical launched by OGFST and achieves national distribution (1976-1980) |
| October 1976 | Organic Growers' Association of Western Australia (OGAWA) (now Organic Association of Western Australia) |
| 1977 | Canberra Organic Growers Society Inc. (COGS) founded |
| July 1978 | Diggers Club, now 'Australia's largest garden club' |

## Fourth Wave – Certifiers – 1980s onwards

| Date | Event |
|---|---|
| 26 April 1986 | Chernobyl nuclear accident |
| 19 May 1987 | National Association for Sustainable Agriculture Australia Ltd (NASAA) registered |
| 22 February 1988 | Biological Farmers of Australia (BFA) registered (deregistered 26 March 2013) (now Australian Organic) |
| May 1991 | Acres Australia founded, 'The national newspaper of sustainable agriculture' |
| 1992 | The National Standard for Organic and Bio-Dynamic Produce implemented as the Australian Export Standard by AQIS (Australian Quarantine Inspection Service) for products labeled organic or bio-dynamic. |
| 1998 | Organic Federation of Australia (OFA) founded |

| 2000 | Australia reports more certified organic agricultural hectares than any other country in the first published global tally of organic agricultural hectares |
| --- | --- |
| 20-27 September 2005 | The 15th Organic World Congress of the International Federation of Organic Agriculture Movements (IFOAM), Adelaide, SA |
| June 2006 | Journal of Organic Systems, first issue |
| 13 September 2009 | All bread-making flour in Australia to have folic acid added from this date; organic flour is exempted. |
| 9 October 2009 | Australian Standard for Organic and Biodynamic Products, 6000-2009 a voluntary standard released by Standards Australia. |
| May 2010 | BFA's Organic School Gardens Program founded |
| 4 October 2011 | Australia's Andre Leu, chair of the OFA, is elected President of the International Federation of Organic Agriculture Movements (IFOAM) at the General Assembly, Korea |

*Source: The Author, 2013.*

lectures was 'The Farm as a Biological Organism' (Paull 2011b). Just months later the world was at war – and that left nothing untouched, including the British taste for Germanic ideas.

Northbourne published *Look to the Land* in May 1940. In the book he adopts Steiner's and Pfeiffer's concept of 'the farm is an organism' and from there coins the term 'organic farming'. He lays the grounds for a contest of 'organic versus chemical farming'. Northbourne's manifesto of organic farming provided a framing with which to consider food and agriculture, and fresh terminology with which to discuss it.

The Australian Organic Farming and Gardening Society (AOFGS) was founded in 1944 in Sydney. It was the world's first association to style itself as an 'organic' association (and, for example, it predated the UK's Soil Association by two years). The AOFGS was frustrated by wartime shortages of paper and the prevailing restrictions stymied their plans for immediately launching a periodical. Their quarterly periodical, the Organic Farming Digest, finally appeared in April 1946 as wartime paper restrictions were eased. The first edition included a Foreword by the Labor Premier of New South Wales, William McKell, who stated that "Exploitation of their land and its produce has utterly destroyed civilizations in the past and if the present civilisation is to avoid its own destruction in the future, then society must give heed to maintaining the fertility and productivity of the soil ... I can give an assurance that this important aspect of our national welfare will never be neglected" (1946, pp.1-2); McKell resigned the following year.

The Organic Farming Digest (OFD) was published for just shy of a decade. The first issue set the style for all subsequent issues in which there was a mixture of articles by Australian, British, American, and occasionally other authors. Over its publication history the Digest included articles sourced from Australia (N=177), UK (N=99), USA (N=82), South Africa (N=7), New Zealand (N=6), Germany (N=2), and Denmark (N=1) (Paull, 2008). The AOFGS published the world's first set of principles of organic agriculture in 1952 – these ten principles predate IFOAM's set of principles of organic agriculture by more than half a century (AOFGS principles are reprinted in Paull 2008).

It is uncertain from whence the AOFGS took up the term 'organic farming', and there are two potential candidates. Northbourne had published Look to the Land in 1940 in which he coined the term 'organic farming'; there were issues by Dent and Basis Books and both appear to have been distributed in Australia. In the USA, publishing entrepreneur Jerome Rodale promptly adopted Northbourne's terminology and published Organic Farming and Gardening in 1942. Whether the AOFGS took up the term 'organic' from Northbourne directly or via Rodale's periodical is undetermined, however one or both of these potential sources are the primary candidates, given the temporal proximity of the coinage of the term in 1940 to the founding of the AOFGS in 1944. The original name of Rodale's periodical is nested in the AOFGS name, and an article by Rodale (1946) appeared in the first issue of the OFD, however Rodale was, in that first Digest, described as the editor of Organic Gardening. What is clear is that there was an active interchange of ideas within the Anglophone world at the time.

In the final two years of the Digest publication was erratic with just a single issue in each of the years of 1953 and 1954. As post-war costs rose, the cover price had soared from 6d for the first issue, later to 1/-, and then to 2/- for the 29th and final issue. The demise of the AOFGS was blamed on lack of funds – the AOFGS had failed to establish a viable financial footing for itself. Nevertheless the Executive Officers declared that "The Society has always operated under a financial handicap, and for this reason the Digest fell short in some respects. However the principles of organic farming have been sufficiently publicised for the work to continue, and the supporters of the organic movement can best promote it by their own example of wise land use" (1954, p. 1). They reported with some justified satisfaction that: "Although the termination of this magazine will be regretted by many, there is solace in the fact that it has performed a service in publicising organic farming principles in Australia" (1954, p. 1). The AOFGS found what some others have since then similarly experienced: "strange as it may seem today, no support was given to the Society by horticultural societies" (1954, p.1).

The Living Soil Association of Tasmania (LSAT) was founded at a public meeting in Hobart in August 1946. The UK's Soil Association had been founded in May of that year and the LSAT was the first foreign society to affiliate with it. The president of the LSAT, Henry Shoobridge (1874-1963), actively recruited members from the outset. Shoobridge a long-established and successful hops grower from Bushy Park, an hour's drive up the Derwent River from Hobart. Shoobridge was well connected and active in civil society. He was a lay preacher for the Methodist Church and a member of the Masonic Lodge. The Secretary of the LSAT wrote that "Our President, Mr. H.W. Shoobridge who is one of the Founders of the Soil Association in England is spending considerable time in personally interviewing interested persons" with the objective of recruiting members (Bayles 1946a, p.1). In another letter Bayles (1946d) wrote that "our President (Mr. H.W. Shoobridge of ☐ Bushy Park☐) is one of the Founders of The Soil Association in England and is keen in widening the interest of interested people in Tasmania". A member was urged to "keep in touch with this office regarding any experiments in organic farming you may be undertaking on your property" (Bayles 1947c).

The LSAT adopted the AOFGS's Organic Farming Digest as its own official publication for members, thus relieving itself of the onus of publishing its own periodical. Membership peaked at 274 individuals (in 1952) (Paull 2009). The LSAT recruited from the outset nine farming associations as members of the LSAT Council.

In contrast to the experience of the AOFGS which failed to attract the support of agricultural societies, the corporate members of the LSAT Council were a veritable who's who of the Tasmanian agricultural landscape. The LSAT Council listed as members: the Tasmanian Farmers Stockowners and Orchardists Association, the State Fruit Board, Stone and Berry Fruits Board, the Royal Agricultural Society of Tasmania, the Horticultural Society of Tasmania, the Upper Derwent Farm Home and Garden Society, and the Tasmanian Farmers Federation. Shoobridge sought broad societal support for the LSAT and he recruited several women's groups to the LSAT Council: the Country Women's Association and the Tasmanian Council for Mother and Child, as well as several government entities: the Education Department of Tasmania, and the Hobart City Council. In its push for social and civic inclusion, the LSAT was more successful than any organics advocacy group before or since.

An aim of the LSAT was to affiliate with the four like-minded organisations of Britain and Australasia:

> We have applied for affiliation with the Soil Association (England), The New Zealand Humic Compost Society, The Victorian Compost Society, and the Australian Organic Farming and Gardening Society (NSW). It is hoped that we will be therefore able to keep in close contact with these kindred bodies and exchange information etc. (Bayles 1947, p. 1).

Part of the vision of the LSAT was to promote organic food growing and consumption within schools. Junior members and Junior groups were catered for within the LSAT membership structure and 'Mother Earth Enquiry Centres' were proposed (LSAT 1947, p.1).

The AOFGS's Digest had been adopted by the LSAT, but the faltering and finally the demise of the Digest was a serious blow to the LSAT because these quarterly publications were the most tangible benefit offered to LSAT members. In 1953 only a single Digest appeared, likewise in 1954, and none thereafter. The LSAT was by this time in terminal decline. Shoobridge was 72 years old when he launched the LSAT, he was still active in 1960 delivering speeches as President of the LSAT at the age of 85 years, but that is the last trace of the LSAT identified by the present author. It seems that there was no succession plan and that the society followed its founder to the grave.

In 1959, Lady Eve Balfour, founder and president of the UK's Soil Association did a great service for Australia's fledgling organics sector by assembling a national overview. It was a service that the sector itself was incapable of managing, given that the AOFGS was by then dead, the LSAT was in terminal decline, and the only other kindred association, the Victorian Compost Society, founded in October 1945, had a regional and a narrower mission.

Balfour had the luxury of a year dedicated to living off the organics sector as she arrived armed with a list of 119 Soil Association members in Australia. She had no budget for travel and accommodation, and the plan was to rely on the hospitality of the Australians. She visited all six states of Australia. She had the ambition to turn her Antipodean adventures into a book, and the royalties would have come in handy, but no publisher took up the project and her account was instead serialised over twelve issues of Mother Earth, the Soil Association's journal (Paull 2011f).

In NSW, Balfour visited Bob Williams "who has a very remarkable Bio-Dynamic demonstration garden, where ... he grows all the herbs required for the various BD preparations ... and he is the principal supplier for BD farmers and gardeners throughout the whole of Australia and New Zealand" (Balfour 1960, p. 397).

Balfour stayed with grazier Colonel Harold White in Guyra, NSW, on his 2,100 hectare property. White had played a prominent role in the AOFGS and was the most prolific of the Australian authors publishing in the Organic Digest. Balfour described White as "one of our earliest Soil Association members" (Balfour 1960, p. 409).

In Tasmania, Balfour met Henry Shoobridge, describing him as "President of the Living Soil Association of Tasmania and our oldest Tasmanian member" (Balfour 1959a, p.702). Shoobridge was an octogenarian at the time and he had been a Soil Association member since the outset. In Melbourne, the Victorian Compost Society (VCS) was still active and Balfour delivered an address to members at a monthly meeting (Balfour 1959b). Also in Victoria, Balfour visited Alex Podolinski ('Pottalinski' in her account) at Wandin. She wrote of "a most remarkable small farm ... I don't think there are very many Bio-Dynamic farms in Australia, but I have never seen a more convincing demonstration of what this, or indeed any other, system of organic farming can achieve" (Balfour, 1959b, p.47).

In Queensland Balfour stayed at Alice Berry's 16,000 hectare sheep station. In Adelaide she stayed with Professor Stanton Hicks who had co-authored a book, Life from the Soil, along with Harold White, in which they lamented the prevailing institutional resistance to the organics message: "here in Australia, the universities and Departments of Agriculture have neglected it, while boosting fertilisers in season and out of season. Indeed, one professor suggested a campaign against the advocates of organic farming before a gathering of [CSIRO] people and was applauded" (White and Hicks 1953, p.95).

Balfour claimed that she had recruited one hundred new Australian members to the UK's Soil Association. That would have been a much needed boost for the Soil Association which was itself in a state of decline at this time (Reed 2010). The visit does not otherwise appear to have borne immediate fruit. Rachel Carson's book, Silent Spring, was still a few years into the future and when it appeared in 1962 it gave the global organics movement a much needed impetus (Paull 2013).

## Third Wave – Disseminators – 1960s and 1970s

The Third Wave of organics advocacy can be anchored from Carson's Silent Spring (1962). The Third Wave ushered in a fresh generation of advocates, organisations, novel promotional activities, a popular and nationally distributed book on organic gardening, a push for organic standards, and a nationally-distributed, regular, attractive, and commercially-successful glossy organics magazine (Table 3.2).

After earlier organics developments in NSW, Victoria, Western Australia and Tasmania, now, in this Third Wave of Australia's organic movement, South Australia comes to the fore. First there was a group of SA members of the UK's Soil Association who came together as the Soil Association (South Australian Branch) (SASAB). It was a sensible and perhaps obvious response to 'the tyranny of distance' that Australians faced along with the shared geography and interests of these members – but no other state took up the example of the South Australians. The SASAB of 1965 grew to become, in 1975, an independent entity, the Soil Association of South Australia (SASA). Most recently, with the decline of membership, momentum and funds, along with the questioning of the ongoing relevance, salience and appeal of the 'Soil Association' name, the SASA has retreated back to branch status, but this time (September 2009) to be the South Australian branch the Organic Federation of Australia (OFASA). Time will tell if this is an innovative way of extending the reach of the OFA or is merely prolonging the demise of the SASA.

The Organic Food Movement (OFM) was active in South Australia as an independent entity from at least October 1972 (and most probably somewhat earlier). In any event, Australia's OFM predates the International Federation of Organic Agriculture Movements (IFOAM) which was founded in Paris in November 1972 (Paull 2010). There was no representation from Australia at the founding of IFOAM and it is a fair speculation that none was sought. The OFA 'amalgamated' to become an 'autonomous section' of the SASAB (Windram 1972, 1973) "to gain recognition, to overcome obstruction to organic farming and marketing" (Windram 1972, p.1). The OFM was incorporated in Adelaide in June 1975: "To promote the production and distribution of organically grown food" (Martin 1975, p. 2).

The first Australian organics book to achieve national distribution was Organic Gardening by Adelaide Hills organic strawberry grower, Audrey Windram (1975). The Rigby Instant Books were a sensation in their day. According to Vanessa Berry (2012): "In the 1970s in Australia, the equivalent of the internet was the Rigby Instant Book … They cost 25-35c … The Instant Books were published by Rigby Ltd, which operated from South Australia with offices in London and New Zealand. The American equivalent of these books, Dell Purse Books, was the same size and shape". The copyright page of Organic Gardening declared that: "The author has donated the greater part of the royalties for this book to the Organic Food Movement". Organic Gardening was number A75204 in the series. Rigby Instant Books were a winning combination in their day bringing together low price, fit-in-your-pocket dimensions, a manageable 64 pages, a breadth of practical titles, ready availability, and practical accessible information.

The Organic Food Movement established a Standards Committee and sought to adopt and adapt the standards of the UK's Soil Association along with their trade mark. The OFM pioneered the development of standards however it appears that this aspect of the OFM did not come to fruition.

The Henry Doubleday Research Association of Australia (HDRAA) was founded in Sydney in 1970. Lawrence Hills had established the Henry Doubleday Research Association (HDRA) in the UK as an association in 1954 and as a charity in 1958 (Hills 1989). He appropriated the name of an obscure and long-dead horticulturist Henry Doubleday (1810-1902) and rather oddly Australia's HDRAA followed this lead. The HDRAA continues to publish a quarterly newsletter Natural Growing for its members. It has not followed the lead of its parent, the HDRA, which adopted the name 'Garden Organic' and in this guise it is the world's largest organic gardening association (claiming 40,000 supporters) and offers visitors a ten acre organics experience at its ticketed-entry organics display garden at Ryton, England. Australia's Diggers Club, founded in 1978, claims to be 'Australia's largest garden club', it promotes organic gardening and its own seeds and is the closest Australia has to Britain's Garden Organic – and it is an image of what HDRAA might have become had it followed the lead of its UK parent.

The Bio-Dynamic Research Institute (BDRI) was registered in 1967 by Alex Podolinsky. The following year, BDRI appropriated Europe's Demeter logo, registering it as its own trademark despite the logo having been used for decades in Europe. The BDRI is not associated with the biodynamic certifying agency Demeter International. Bio-Dynamic Gardeners Association Inc. (BDGAI) was registered in 1991. According to the BDRI website, demeter.org.au, "In 1953 the Bio-Dynamic Agricultural Association of Australia (BDAAA) was founded" although this claim could not be verified by the present author and no trace of the BDAAA was located in ASIC records.

The Organic Gardening and Farming Society of Tasmania (OGFST) has some claim to be Australia's most successful organics advocacy society. It was established at a meeting in 1972 at the University of Tasmania and was active until its wind up in 2009, making it one of the most long-lived of Australia's organics societies. At its peak the OGFST had about 1000 members, it proliferated across the state of Tasmania, with 22 branches recorded, and even to the mainland, with branches in Canberra, Ballarat, and Victoria. The OGFST produced a blizzard of publications – many branches issued their own newsletters, there was the OGFST members' newsletter Grapevine, there were various pamphlets and booklets including the Organic Gardener's Diary and Beginning Your Organic Garden – Nature's Way to Grow.

The OGFST produced the first nationally successful organic magazine. It began as The Organic Gardener and Farmer in 1976, and evolved to become Organic Growing. There were 192 issues with the final Organic Growing appearing in 1994. With print runs of 11,000 copies, Organic Growing carried the organics message far beyond its membership and to all states of Australia (Stevenson 2009). The magazine relied on the enthusiasm and labour of a bevy of volunteers.

The OGFST presented thirty two Organic Festivals, including one children's festival, in the years 1975 to 2005 (Stevenson 2009). The OGFST set up a permanent venue for its festivals on acreage outside of the picturesque town of Penguin, North West Tasmania. Over the three decades of OGFST Organic Festivals there had been showbags and pony rides for children, talks at schools, and horticultural demonstrations, but at the final AGM in 2009 one octogenarian member confided to the present author: "We are now too old to dig". It was a reminder that civic associations, even one as successful as the OGFST, need to continuously pay attention to recruitment and succession.

In this Third Wave, the age of disseminators, there were other regional grass-roots associations of organics enthusiasts. These included the Organic Growers' Association of Western Australia (OGAWA) (now the Organic Association of Western Australia) established in 1976, and the Canberra Organic Growers Society Inc. (COGS) established in 1977. The OGAWA produced The Organic Grower (1981-2006) and their periodical is now WA Organic Life. The COGS produced their Newsletter (1982-1992), which was replaced by the COGS Quarterly (1993-1999) and now Canberra Organic (1999 to the present).

Groups of enthusiasts proliferated in this era of the Third Wave of organics in Australia. Other kindred groups included: the Brisbane Organic Growers' Group, the Organic Farming and Gardening Society of Victoria, the Organic Growers Association (NSW), the Healthy Soil Association

(Qld), and the Natural Health Society of Australia. This era of disseminators was a time of amateur enthusiasts, of optimism, of volunteering and of civic engagement, and was quite different from the next era of certifiers which ushered in organic corporates, standards, certification, labeling, monetisation, along with an identifiable, quantifiable 'organics sector'.

## Fourth Wave – Certifiers – 1980s to the present

The Fourth Wave of organics in Australia was ushered in by a global event, Chernobyl. The explosion of the nuclear reactor at Chernobyl, Ukraine, on 26th April 1986 dramatically focused the world's attention on food safety issues. Radioactive sheep in Wales and radioactive reindeer in Sweden were destroyed. In Australia imports of food from Europe were subject to bans, restrictions, and radioactivity testing.

This Fourth Wave of Australian organic development brings us to the present age, the age of certifiers, certification, standards, labeling, logos, corporatisation, and the monetisation of the organics project. The Fourth Wave has been the beneficiary of all those advocates, enthusiasts, visionaries, entrepreneurs, and innovators who went before, who were the grass roots of the movement and who explored a myriad of ways and means of advancing the organics project. The forebears of the Fourth Wave drew heavily on volunteers who struggled to move their advocacy onto a sound and enduring financial footing. This Fourth Wave has witnessed the founding of the first certifiers, the monetisation of the process, the setting of standards, the export of certified products and in general the corporatisation of the organics project (Table 3.2).

The National Association for Sustainable Agriculture Australia (NASAA) was incorporated in 1987 and soon after Biological Farmers of Australia (BFA) was registered in 1988. Although there are now a handful of other organics certifiers in Australia it is NASAA (based in the Adelaide suburb of Stirling) and BFA (based in the Brisbane suburb of Chermside) that were the trailblazers in developing standards and certifying to those standards, and, having staked out the ground in the 1980s, they remain the leaders in the field. At the time of their founding, neither NASAA nor BFA had the perspicacity to incorporate 'organic' in their names – or perhaps it was a lack of courage or commitment or clarity of vision? Whatever, NASAA has persisted to this day with its ungainly acronym, while BFA now uses its entity Australian Certified Organic Ltd. (ACO) (registered in 2002) as its certification arm and 'Australian Certified Organic' as its certification tag on certified products, and it has changed its own name to Australian Organic Ltd (registered in 2009).

There are currently two national organics periodicals and both began during this Fourth Wave period. Acres Australia began in 1991 and tags itself as 'The national newspaper of sustainable agriculture'. It released its milestone one hundredth issue in March 2013. The glossy magazine Organic Gardener has been published by the Australian Broadcasting Commission (ABC) out of the Sydney suburb of Alexandria since 1998. The ABC's Organic Gardener appears to have taken up the niche developed over the previous two decades by the OGFST's Organic Growing magazine. In settling into this niche, Organic Gardener has followed the name changing 'tradition' of the previous national organics magazines, Organic Farming Digest (1946-1955) and Organic Growing (1976-1994). Before settling on its current title of Organic Gardener it has variously been titled as: Gardening Australia's the organic gardener; Gardening Australia organic gardener; Gardening Australia the organic gardener; Gardening Australia's organic gardener; and Gardening Australia. The good news is that while Organic Gardener began life as a quarterly (like its predecessor Organic Growing) since 2007 it has been published every two months.

The Organic Federation of Australia (OFA) was founded in 1998 and claims to be 'the peak body for the organic industry in Australia' (OFA 2013). The raison d'être of the OFA is to unite the 'organic industry' and to represent 'the interests of Australia's organic and biodynamic producers to industry and governments at the local, State and Federal level'. These objectives remain aspirational. After 15 years, the scorecard shows that (a) a unity under the OFA has failed to materialise, only two certifiers (OFC and NASAA) out of seven Australian organic certifiers are members and nor are most producers members, and (b) Australian government support for organics is almost non-existent.

The OFA claims as an achievement the Australian Standard for Organic and Biodynamic Products, 6000-2009 (SA 2009). Others may view this as a backward and retrograde step, and the voluntary standard released by Standards Australia lives behind a pay-wall owned and operated by SAI Global. The copyright of this standard belongs to Standards Australia, and not the organics sector. The consequence of this is the creation of a new business selling this organics standard of which SAI Global appears to be the sole beneficiary. While the planet is moving towards openness, OFA has taken this step in the opposite direction, denying consumers what they are surely entitled to know — for what they are paying when they pay a premium for organic. The OFA has 'achieved' a standard that is competing with the open-access National Standard for Organic and Bio-Dynamic Produce (OIECC 2009) which remains freely available at http://www.daff.gov.au to download.

The OFA appears to be proving over again what earlier organics groups in Australia have demonstrated repeatedly, namely, that a business model relying on membership subscriptions is a tenuous model. The chair of the OFA, Andre Leu, was elected to the presidency of IFOAM in 2011, and this must be chalked up as an achievement of the OFA and hopefully in time this can deliver tangible benefits for Australia.

IFOAM's show-case event, the World Congress of Organic Agriculture, was held in Adelaide in 2005. The Adelaide World Congress was a first for the Southern Hemisphere and attracted leading organics researchers and advocates to Australia, many for the first time. Adelaide was an ideal choice for this triennial event. Adelaide was founded as a utopian enterprise, planned as a utopian city by visionaries, and has been described as a 'Paradise of Dissent' and a 'Radical Utopia' (Pike 1967).

Along with the World Congress came the 8th International IFOAM Organic Viticulture and Wine Conference, the International Scientific Conference on Organic Agriculture, and the IFOAM General Assembly.

The Journal of Organic Systems (JOS) was established as a joint Australian/New Zealand venture in 2006. JOS is a free, open-access, peer-reviewed academic journal dedicated to publishing organics research. JOS has published 12 issues to date, with research published from around the world. It is the world's leading open-access organics journal, and it is the first to offer readers the option of an e-book version.

## Conclusion

Global figures for the organics movement have been published annually since 2000. They reveal that Australia has been in the lead position in terms of certified organic agriculture hectares from the outset of this dataset. In 2000 Australia reported 1,736,000 hectares (World total = 7.5 million ha.) and in 2013 reported 12,001,724 ha. (World total = 37,245,686 ha.) (Willer et al. 2013; Willer and Yussefi 2000). Of Australia's agricultural land, 2.93 per cent is managed organically (Falklands/Malvinas = 35.94 per cent; World = 0.86 per cent). There are 2129 Australian organic producers (World: 1,798,359). Annual organic sales are reported as AU$1.178 billion (€942,000; World = AU$59.779 b, €47.805 b) (Willer et al. 2013). Unlike many other countries, Australia is yet to report organic wild hectares, organic forest hectares, or organic aquaculture (Paull 2011d).

There continues to be little support from government in Australia for the organics sector. The CSIRO (Commonwealth Scientific and Industrial Research Organisation), Australia's national government entity for scientific research, has not been a useful ally in developing the organics sector, and despite there being zero consumer demand for GMOs, it is working contrarily to develop and commercialise genetically modified crops.

There is some ancient wisdom that declares "Where there is no vision, the people perish" (King James Bible, Proverbs, 29:18,). Bhutan has boldly declared a vision of being 100 per cent organic, and others countries and states have declared lesser goals. Meanwhile, despite its long history of engagement with organics advocacy, as revealed in this chapter, Australia has no declared plan, vision or 'roadmap' for growing beyond its current status as a minor niche player. The World of Organic Agriculture Statistics and Emerging Trends has reported no change in Australia's organic hectares for the past three annual issues (Willer and Kilcher 2011, 2012; Willer et al. 2013), and what seems likely is that the momentum of the growth of Australia's area under organic management has stalled.

The First Wave of organics in Australia witnessed the take up of Rudolf Steiner's call for a differentiated agriculture. The Second Wave witnessed the first associations for the public advocacy of organic agriculture. The Third Wave saw the national dissemination of organics know-how and rationale. The Fourth Wave witnessed the corporatisation of the organics sector along with the formalities of standards, certifiers, labels and logos. A Fifth Wave is now perhaps overdue to pick up the efforts of ten decades of advocacy and to carry them forward to the bigger, better, and bolder future that was always imagined.

# 4

# Profiling the Australian organic market

*Andrew Monk*

## Introduction

This chapter discusses the nature of the Australian organic industry, its regulatory functions and the growth and development of its markets since its formal inception in the 1980s. The Australian organic industry manifests a mature mixture of government and industry self-regulatory arrangements, where there are laws outlining export requirements, domestic consumer protection laws, and a market driven industry association structure overseeing standards and certification functions for domestically produced and traded products.

Unlike other markets such as the US, EU and even Canada, Japan and Korea, there are no specific domestic laws in Australia in relation to government mandated standards on organics or on the mandatory requirement of certification by the organic industry. Nevertheless, requirements for organic production and marketing in Australia mirror international trends of organic standards and certification. Also, considering the advantages that Australia has in its natural production and a well-resourced and networked industry association structure, the organic industry has maintained some of the highest and most stringent standards in the world. This is in the face of having relatively little government support for the organic sector, especially in the areas of research and development, and even in industry regulatory functions, which is largely funded by the organic industry itself.

Even with very small differences in organic production standards around the globe, many overseas governments have regrettably chosen to establish their own, slightly differing, market regulatory requirements (certification and related accreditation laws pertaining to organic) which continue to create challenges for Australian organic producers and exporters wishing to enter overseas markets. This has added costs as well as sometimes complexities in regulatory compliance for Australian exporters and their suppliers. It has resulted in the need for differing paperwork and sometimes differing production requirements for different markets – albeit with almost identical on the ground production requirements.

Domestically the organic industry has experienced solid growth over the past few years, most particularly noted by the more committed involvement of the larger retailers. Hence the types of products available, and the producers and processors entering this industry to supply these products have evolved to match consumers' needs and their different demands and expectations for organic products. Nonetheless the market remains very diverse and is also flourishing with the roll out of independent organic stores, direct to home retail, and even farmers' markets, offering organic producers and marketers numerous options in the Australian marketplace. One of the biggest barriers to consumers buying organic is availability, convenience and ease of access which remains a challenge for marketers. However the presence of large retailers is starting to see this barrier being marginalised. Cost, value and trust issues for consumers remain the other predominant barriers to purchasing organic products.

There are a number of promising sectors for Australia in the years ahead. More recently red meat, which is beef and lamb and also the wine sector have exhibited significant growth. However, in terms of supply the red meat sector in particular continues to struggle with sufficiency and consistency in the supply of the primary product, given growing demands domestically and for export. This unmet demand signals future growth in the number of operators and obviously in output which is envisaged for this sector at the primary producer end. Some sectors like horticulture have in many respects seen a flooding of markets, in which the small to medium operators have been displaced by the larger retail chains which enjoy economies of scale. Again sufficiency and consistency of supply seems to be an issue. The horticulture sector in particular is likely to see more growth in the future, but possibly not in the number of operators. This growth is envisaged in expansion of existing farming units or overall production volume growth as some medium and larger sized operators are new entrants in this industry.

Other sectors have struggled with lack of sufficient primary products. This is true for the grain sector, which has benefited from the recent breaking of drought conditions in some areas of the country, albeit with the challenge of managing weed and pests during these better seasonal periods. This also highlights the need for more organic specific research and development in Australia for such sectors where the production requirements differ markedly from conventional primary production practices. Food manufacturers generally are struggling with a high Australian dollar, and higher relative costs of production for value adding. This may continue without foreseeable end. Nonetheless with Australia's reputation as a country with natural production advantages, along with an international reputation for stringency in relation to regulatory functions and transparency, there are opportunities of significant growth for export of organic products in the years ahead. Similarly, working from a low base which is approximately one per cent of retail value, there is considerable potential upside for the organic market domestically, as compared to matured international markets of the US or EU. In these matured markets, organic products constitute approximately two to three per cent of the total retail value. These figures highlight future opportunities of growth in the organic sector for Australian producers and suppliers.

## The regulatory environment in Australia

The organic industry and organic consumer movement have had a formal market presence in Australia since the 1980s with the establishment at that point of industry standards and related certification programs. By the early 1990s the Australian Government became involved with the introduction of organic produce Export Orders, seeing a legally binding definition for organic in the National Standard for organic and biodynamic products, and related requirements for certification for export (organic became a 'prescribed good' for export). This was in reaction to the establishment of legislation in the EU outlining requirements for the production and marketing of organic products.

This move ensured that Australian producers and exporters were able to maintain market access to those countries within the EU, which until then had been otherwise regulated via industry self-regulatory structures or in some cases EU member state regulations only. This move also harmonized the organic industry in Australia in terms of standards applied in the field and along the production chain. The likes of Japan and the US followed suit with binding import requirements and related domestic market regulation in the late 1990s and early 2000s.

Two of the main associations that to this day dominate the Australian standards and certification landscape are Australian Organic Ltd (formerly BFA Ltd) and the National Association for Sustainable Agriculture, Australia (NASAA Ltd). These two associations maintain their own standards and certification systems and the industry as a whole is harmonized by a peak council binding the certification programs, via industry agreed structures and agreements. This combined with a more recent (2009) presence of a Standards Australia organic standard, which while voluntary, has assisted in setting an agreed baseline or benchmark requirement for organic production and marketing in Australia.

Ultimately at the domestic level, Australia manifests industry self-regulatory arrangements which are largely driven by industry and supported by the marketplace, and further underpinned by Australian government consumer protection laws. This has resulted among other things in one significant fraud conviction, and several other investigations based on sales of products claiming certified organic status to an industry certifier's standards. Australia awaits a precedent in law in relation to uncertified products which claim organic status. In this vacuum, and in fact because of it, larger retailers in particular have supported the industry programs of standards and certification to eliminate this risk and to bolster consumer confidence, while also protecting their own trading interests. This to a great measure works in the marketplace, although it is less clear whether at the more informal market levels comprising smaller retailers, farmers markets and direct marketed schemes, consumers are as protected under law, given the more regular presence of uncertified organic marketing claims.

This industry exhibits significant diversity across its supply chains, and considerable competition for service at the certification agency level, delivering a relatively low cost regulatory burden for the industry and in turn to consumers relative to other more government regulated markets. It will remain to be seen whether in the years ahead this unique hybrid Australian approach to market regulation will stand the industry in good stead in terms of market and consumer regulatory protection. Whatever the case the ongoing complexities facing Australian exporters do not appear like they will be going away any time soon, in terms of ongoing costs and burdens for those with this market focus – something that cannot be solved by Australian producers nor regulatory authorities in any significant measure. In this context leading certification agencies, acting on behalf of their clients, have achieved international market access recognition by directly applying and achieving accreditation and recognition into these markets.

## The Australian market and its place internationally

The Australian organic market has continued with sustained growth through 2010–2012, building on prior years of growth, in the face of margin declines as the industry has matured and moved more 'mainstream'. The total value of the organic industry in Australia was estimated at $US 1.276 billion in 2012. The average retail growth projection, taking into account the historical growth in the past few years is approximately 10 to 15 per cent. While this is a conservative estimate, it is clear that some sectors like dairy, horticulture and meat, might exceed these growth estimates. Total farm-gate value of certified organic products in Australia in 2011 was conservatively estimated at $US 300,637,412 and total farm turnover from farms with certified organic status at $US 432,211,807. Internationally there has been growth year on year of organics. Growth has ranged from 11 to just over 2 per cent between 2007 and 2010 (Euromonitor 2011b). The UK market appears one of the exceptions to this ongoing significant growth in the face of suppressed economic conditions. The total value of the organic marketplace internationally was estimated at $US59 billion in 2010.

The Australian organic industry continues to command a relatively small percentage of total market value with a sectoral range estimate of between 0.8 to 1.2 per cent, which is lower than the growth rate for conventional produce. It is envisaged that in the coming decade, organic sales as a share of retail value might reflect those currently in EU and US, which is between 2 and 3 per cent. However, for this to occur, several supply chain constraints would need to be addressed. Whilst it is more difficult to track and measure the value of imported organic products in Australia, it is estimated to be in excess of $220m in 2012. This figure has increased as a consequence of supermarkets attempting to fulfil demand, with importers and consumers benefiting from a strong Australian dollar. Imports include a rising trade in processed goods from the EU and the US as well as from the Asia-Pacific region, along with base ingredients including milled grains and livestock feeds, essential oils and dairy powders. These imported items are used by Australian manufacturers who are unable to procure adequate local supplies.

Exports of organics have remained generally suppressed in

comparison with figures from the early 2000s, with the exception of successful examples in meat and dairy. Exports are estimated to be 10 per cent of Australia's industry value, which was approximately $126m in 2012. In the coming years, a falling Australian dollar and increased supply capacity offer opportunities for this industry to supply into established international markets of the developed world, as well as to expatriate communities and middle class consumers in developing economies.

Organic sales are increasingly becoming mainstream in Australia. In 2012, 92 per cent of organic sales were through store-based retailing. Three out of four organic purchase experiences are now at major supermarket chains, underscoring an ongoing 'mainstreaming' of organic products, even while independent retailers and other retail outlets continue to experience growth and in some instances very high growth. The ability to develop domestic production to meet this demand continues to be a key challenge for future growth of the Australian organic industry, with some sectors like processing for freezing and other value adding, reporting inconsistent or unavailable supply of raw ingredients to deliver on known demands. This continues to highlight opportunities for those interested in importing into the Australian marketplace or import substituting by Australian suppliers. Figure 4.1 (next page) depicts the retails sales growth of organic products in Australia in the past two decades.

## Australian organic primary industry sectors

The estimated farm-gate value of organic production has grown 16 per cent per annum since 2009 to $300,637,412 in 2011, which is a 34.67 per cent increase over two years. This increase was achieved in the face of pressures associated with lower pricing. Hence the additional volume growth is not reflected in these growth figures. The breaking of drought in Australia in 2010-11 has been of significant benefit to many primary producers in affected areas. However, it has been a mixed blessing, also bringing with it challenges of additional weeds, particularly in cereals,

**Figure 4.1: Australian Retail Sales Growth 1990 - 2012**

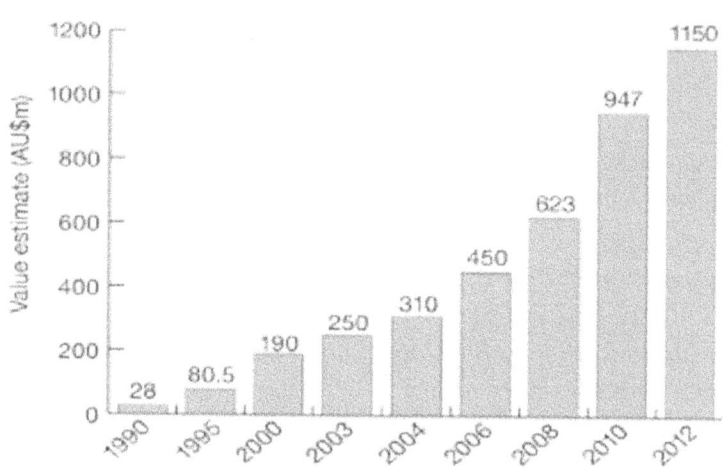

*Source: Australian Organic Market Report (2012)*

and diseases in grape production, which reduced or decimated some harvests.

The organic industry is continuing to mature and the average size of organic farms has increased, highlighting a trend towards professional farming on a larger scale, albeit smaller than the average size for conventional farms in most sectors. This also highlights the expansion of some long-term organic farming families who have purchased additional land and/or farm units in other states to cater for increased demand as multiple retailers move more decisively into the organic market. While mainstreaming and becoming professional, the organic industry remains diverse in terms of operator types and sizes, with the continued success of farmers' markets and direct-marketed products. The number of smaller-sized certified organic operations remains high. As is noted in Figure 4.2, the growth in farm gate value over the past decade has been solid, rising significantly over the past few years – in fact in excess of the growth in overall retail value. This possibly highlights the increased efficiencies in the post farm sector of the industry through to retail, while also highlighting a growth in demand and supply of Australian organic produce.

**Figure 4.2: Australian Farm Gate Sales Growth**

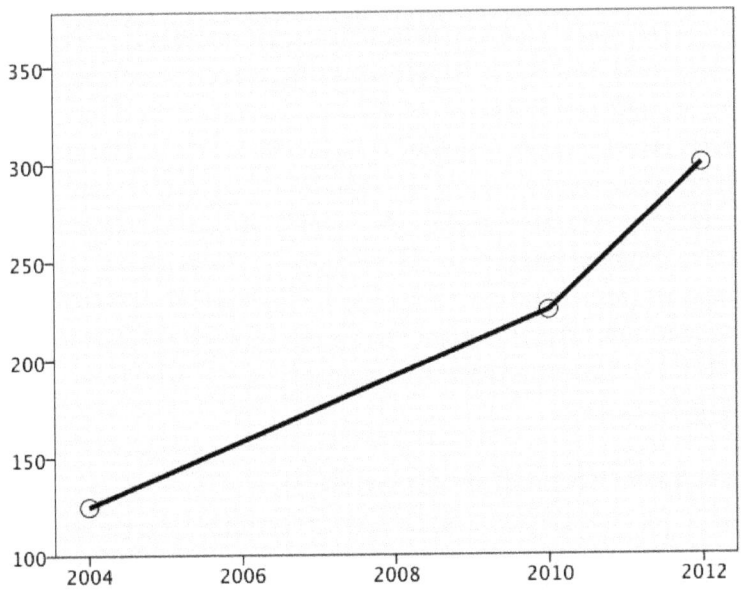

Source: *Australian Organic Market Report (2012)*.

The coordination of organic production and supply chains remains the biggest opportunity and challenge for some sectors; however there are excellent examples where this is now gaining traction and resulting in higher-value products such as snack-size and squeezy pack yoghurts and beef patties. Organic farm inputs continue to grow in product numbers and businesses. In 2012, 187 businesses in Australia had formally registered organic farming inputs, like fertilisers, bio-pesticides, crop management inputs and other approved processing and cleaning products. This industry is incredibly diverse as per industry sectors. In addition it includes small, medium and large size producers who supply to a variety of market outlets, which range from local and direct through to big retail suppliers and exporters. This creates further challenges for this industry in terms of organizing supply chains and delivering to a consistently growing demand.

Horticulture remains the dominant industry by volume and value. However red meat, in particular beef, has risen significantly over the past two years, and now is the largest sector by value in the industry. Apart from red meat and horticulture, dairy is Australia's third most notable and solid growth sector, with much promise ahead for it. The grain sector has been challenged in terms of supplies of primary product. Finally, essential oil and related sectors that supply the cosmetics industry are tipped to grow significantly in the next decade. Consistency and growth in supply of primary product will remain the single greatest challenge for the organic industry in the years ahead. This is necessary in order to instil confidence for the purchasers of organic products, who do so for retail and export markets.

## Certified land area and supply trends

Australia has the largest surface area of certified organic land in the world. In 2011, the Australian Bureau of Statistics (ABS) data noted that 11,199,577.4 hectares was certified from a total property area of 13,637,541.9 hectares, which included non-certified organic areas of operations. The available certification agency data suggests a significantly higher figure of 16.9 million hectares is certified as organic in Australia. An additional reported 253,392 hectares is in current 'pre-certification' phase, with this figure likely to be higher. This represents land area that over 2012 and 2013 would convert into the formal certified market as it completes the conversion period requirements for farmland.

Large tracts of rangeland account for the majority of this large certified land area, namely organic cattle in the Queensland Channel Country (South West Queensland) and rangelands in western New South Wales and South Australia. Queensland has the most organic certified area of Australian states and is the state with the single most certified area of land in the world, along with the highest value of organic agricultural production within Australia. New South Wales accounts for the highest number of organic operations in terms of number of individual certified organic businesses. Victoria harbours the majority of organic dairy production in Australia by value, and along with NSW has the highest activities in the manufacturing and value adding sectors.

Nationally it is estimated that at the end of 2011 there were 2,117 certified organic primary producer operators recorded, some of who may have more than one organic farm holding. According to ABS statistics some 1520 operations were reported as agricultural businesses in the 2011 Agricultural Census data. The disparity here is due to both a number of smaller-scale farms that are not registered with ABNs (Australian Business Numbers) and/or are boutique or informal in nature and therefore do not register on the ABS data tracking system. This is also the reason for the disparity in terms of overall estimates of certified land area.

The growth in new farm as well as farmer (operator) certifications has slowed in the past two years from 2010, down from the average of 5 per cent net growth between 2002 and 2009. These figures do not reflect the ongoing increase in overall average farm size, along with additional farm units being brought into production by existing certified organic farming businesses. It is the latter, i.e. growth in volume on existing operations or by existing operators, that is fuelling the majority of growth currently seen in the organic market.

## Organic consumers

The type of organic consumer in Australia is changing, with more mainstream uptake of organics as major retailers range organic products. Three out of four purchases of organic products were made at a major retailer in 2011, representing a marked shift as more mainstream consumers purchase organic products, which highlights the nature of future volume growth for the organic market sector. According to estimates in 2012, over one million Australians purchase organic foods and beverages on a more than infrequent basis; 65 per cent of consumers surveyed purchase organic food occasionally, which is higher than figures of 40 per cent in 2008 and 60 per cent in 2010. Organic fresh produce is the first entry point for most first time consumers and represents a major spend in the organic shopping basket, while non-alcoholic beverages score the highest in terms of regularity of purchase by organic consumers. The major barriers to purchasing organic products remain price and availability, though both have reduced as barriers in 2012, along with an increase in trust and a significant growth in recognition by consumers of the importance of certification labels on organic products. Changes in availability and convenience, which has largely been brought about by the entry of the major supermarket chains, have led to significant reduction in the barriers to purchase as compared with the statistics of 2010. Additionally, consumers have a high regard for the integrity of Australian products and the Australian certification of organic foods, as compared to the products belonging to some 'other country of origin' products.

The average Australian organic consumer places emphasis on the produce being 'chemical free' in comparison to European consumers whose emphasis is more on environmental attributes as well as high and additional expectations for animal welfare outcomes under organic practices. These differences between the average Australian and European consumer could possibly be owing to a general Australian attitude of living in a cleaner environment, with a lower degree of scandals in relation to animal welfare. Issues associated with 'genetic modification' or 'GE' food products also seem to more prevalent in the EU, albeit many organic consumers expect and perceive organic products to be free of such ingredients. Arguably in the EU there have been more activist based food campaigns which have possibly galvanized a segment of consumers to demand organic as their assurance that their food products are better than just legally approved by their government regulatory authorities. There appears less of this concern by Australian consumers currently, though this could easily change should an Australian specific food scandal arise in future.

The challenge for the organic industry in the years ahead will be to supply the growing demand for organic products, though ongoing growth will likely only result from lower retail prices for organic products, along with greater consistent availability of the main consumer staples. This mix of factors has remained by far the greatest hurdle in relation to significant additional growth for the Australian organic marketplace, its producers and value adders.

## Value adding and market access challenges

In 2012, there were 747 operators certified as value adders (processors) or marketers (wholesalers and exporters) in Australia. Combined with certified farming operations and other certified operations like farm input businesses, the total number of certified operations in 2012 was 3069. The margins for processors and marketers have not been consistently reported while overall turnover in general is up due to higher throughput in 2012. The majority, which is around 62 per cent of operators, are bullish about the future growth prospects of the sector and they envisage that there would be an increase of up to or more than 10 per cent in 2013. Another 33 per cent of operators expect sales to be stagnant in 2013 and only the remaining 5 per cent expect declines of up to 10 per cent or more.

Australian-based certifiers maintained up to eight international country access and market accreditations including for the US, Japan, Canada, the EU, Korea and IFOAM and were active in 12 countries (as part of import/export trade into and out of Australia) with some 85 client operations. This is in addition to the Australian government having an export standard and regulatory scheme in place to achieve 'equivalence' arrangements with importing countries. Such equivalence arrangements between governments, has regrettably been in decline through the 2000s, adding further costs and regulatory burdens and complexities to processors, exporters and importers. The Chinese market, while offering great future potential for Australian businesses, has more recently made trade growth difficult due to regulatory changes that do not recognize existing international organic regulatory systems. Specific market requirements in the US, Japan and Korea, including additional certification requirements, are adding cost and complexity to exporters interested in serving this emerging and significant future market. Australian processors, exporters and importers need to remain careful and aware of such developments and requirements of the importing country prior to making investments in organic production and marketing ventures. For the well informed and prepared, the organic marketplace offers a range of opportunities and value adding options, with a short to medium term outlook of ongoing growth in demand for such products.

## Conclusion

The Australian organic industry in many respects is a paragon of industry self-regulation, which is driven by consumer and market support, largely in the absence of government involvement, financial input or support. There remain a number of opportunities for primary producers and marketers alike in this marketplace, albeit the usual caveats that apply to any open market driven industry. Issues relating to the ebb and flow of over and under supply, are particularly applicable for this smaller and emerging market, which can flood more easily with demand and changing specification requirements. Some of the greatest challenges for the organic industry as a whole and the marketplace are the supply of consistent product to the specifications of exporters, importers and large retail markets. Producers have many choices within these markets, including a healthy and vibrant independent retail sector supplied by existing organic wholesalers, through to direct marketed schemes and home delivery.

The average organic consumer is changing and in Australia is focused on produce being 'chemical free' in comparison to European consumers where there is more emphasis also placed on environmental attributes as well as high and additional expectations for animal welfare outcomes under organic practices. Whatever the case, the new organic consumers in Australia are those that expect their shopping experience to be easy with full access to the current range of non-organic products that they are used to shopping for, while also at a price that is closer to conventional than has possibly been the case in prior years. This in its own right poses challenges for suppliers and retailers alike as they try to match consumers' expectations with producers' and marketers' abilities to be able to deliver products within a price range that is also sustainable to their own businesses. This challenge will remain a major one for the industry in the years ahead, and the pressures cut both ways. Simply put, if the price falls below certain levels, the 'risk premium' often associated with producing organic at the primary level may drive future and some existing suppliers to exit this industry, a move that would benefit neither consumers nor retailers.

Exports from Australia are likely to rise in the coming years, with an expected devaluation of the Australian dollar in the medium to longer term, along with a growing supply base of producers able to supply to market volume and other specifications. Imports will also continue to rise, mainly driven by a mixture of the lack of an internationally competitive food processing industry in Australia for some sectors combined with the overall rise in demand for products simply not produced by Australian businesses. As the market continues to grow and diversify such a situation is inevitable, as it is in overseas markets where some Australian producers and marketers are equally benefiting. Australian organic research and development is now seriously lagging behind its international peers. Whilst Australian producers can to an extent utilize the existing available international research in this field, there will soon be a sizeable gap in industry needs versus research knowledge and know how in Australia. This scenario needs considerable rethink by the industry and the Australian government alike, and measures should be taken to counter it.

The next five years ahead look very promising for the Australian organic industry in relation to tangible and latent demand domestically and a solid international demand for organic products. Additionally, Australia has an efficient producer and market focused industry with growing networks, and in some cases marketing co-operatives, which would place Australia in a good stead in the future. Having said this, the organic industry is facing the risk of not keeping up with the rising demand, and the consequences of other marketing claims and food label attributes, for example 'clean and green' claims, that may in the future dampen or distract some of this opportunity.

# 5

# Consumer preferences for organic food and LOHAS

*Antonio Lobo*

## Introduction

It is obvious that given the choice, people would prefer to purchase organic food which is considered devoid of synthetic pesticides and chemical fertilisers. However, not all consumers actually purchase organic foods. This implies that there are several factors which influence consumer preferences of organic food consumption. Furthermore, although the value of the organic food industry in Victoria (Australia) is in excess of $500 million per year (Lobo et al., 2012) and it is one of the fastest growing food sectors in the Australian food industry (Henryks and Pearson 2010), its identification as a valued customer segment is debatable. Hence it would be useful to discuss why consumers prefer organic food as compared to conventional food and also the factors that encourage or inhibit organic food consumption.

As noted in the previous chapters of this book, not many consumers have a clear understanding of the characteristics of organic food. A study which synthesises previous research findings suggests that the lifestyle and other demographic variables of organic food consumers are diverse (Hughner et al. 2007). This therefore makes it difficult to effectively examine organic consumer segments, their preferences and expectations. Several academics have recommended rigorous research to be carried out relating to consumers' perceptions on health benefits, safety, food quality, motives and assumptions in buying organic food as well as country differences in organic food consumption (Thøgersen,

2010). The purpose of this chapter is to discuss the existing debates relating to consumer preferences in the not so well known world of organic food consumption.

Consumer preferences and attitudes are highly related. According to Michaelidou and Hassan (2010), the attitudes that drive organic food consumption can be well explained by investigating concern for food safety, ethical lifestyle and price. Attitudinal variables and other antecedents of organic food consumption have been dealt with in the second chapter of this book. This section focuses on specific factors that influence consumers' preference in organic food consumption.

This chapter is organised in several sections. Firstly, a general review of previous research on consumer preferences relating to organic food consumption is presented. Secondly, health and safety concerns in organic food consumption are discussed. Next, price sensitivity in organic food consumption is discussed, followed by a discussion on organic labelling and certification. Then there is a discussion on the impact of lack of availability on organic food consumption followed by an explanation of the interesting concept of Lifestyles of Health and Sustainability (LOHAS). Finally this chapter concludes with a summary.

## Consumer preferences relating to organic food consumption

According to previous research, consumer preferences for organic foods are dependent on several factors. They include, product (appearance, sensory appeal, perceived benefits and price (Lockie et al., 2004), government regulations and certification (Athanasios Krystallis and Chryssohoidis, 2005), lifestyle (self-indulgent and variety seeking) (Yang, 2004) and ethnocentrism (preference for domestic products over imported products) (Chryssochoidis et al., 2007).

More recent research also highlights the changes in consumer preferences relating to organic food consumpion. Based on a review study of organic food consumption over the past few decades in the USA, a group of researchers report a significant shift in consumer preferences towards locally grown food (Adams and Salois, 2010). These researchers find compelling evidence that an increasing number of US consumers support locally grown produce regardless of whether they are organic or not. They also highlight the lack of the usual discourses that are centered around organic food, such as small farms, animal welfare, sustainability and community welfare.

According to another study on organic food buyer behaviour in Victoria (Australia), although consumers perceive organic food to be chemical free and healthier than conventionally grown food, price is one of the main barriers to their consumption of organic food (Lobo et al., 2011). Some studies find evidence that the appearance of organic food discourages consumption. The aforementioned study, however, finds that although there is no relationship between the appearance of organic food and its consumption. Victorian consumers do not perceive organic foods as having better taste than conventionally grown foods (Lobo et al., 2011). As such, it can be seen that consumer preferences in organic foods are shaped by several factors some of which are more prominent than others.

According to extant literature, health and safety are the prominent perceived benefits that motivate organic food consumption. Further, high prices and lack of availability usually inhibit organic food consumption. Misinterpretations relating to labelling of organic food and inconsistencies in organic certification procedures tend to confuse organic consumers and this inhibits organic food consumption. These and other factors that determine consumers' preferences for organic food are discussed in the sections to follow.

**Health and safety concerns relating to organic food consumption**

Despite facing difficulties in sourcing for organic food owing to its limited access and relatively high price, many consumers perceive a relationship between organic food consumption and health benefits (Bonti-Ankomah and Yiridoe, 2006). Although several studies suggest that consuming organic food results in some health benefits, other recent studies disconfirm this phenomenon (e.g., Dangour et al., 2010). Hence, the real health benefits associated with the consumption of organic food is highly controversial. Nevertheless, it can be seen that many organic consumers tend to believe that organic foods have more health benefits than conventional foods. Therefore, the consumers are willing to pay a higher price for organic foods than for their counterparts (Gil et al., 2000). Willingness to pay higher prices for organic food will be dealt with in the next section.

In a study of European consumers' perceptions of organic food (consumers were from Italy, the United Kingdom, Hungary and Denmark) a group of researchers demonstrate that concerns regarding food safety influence the organic food consumption of many of the European consumers (Torjusen et al., 2004). According to these researchers, the use of pesticides and food additives are of concern to many European consumers who are also concerned about environmental and health effects of conventional foods. As such, the European consumers prefer organic foods that are believed to be free from pesticides and food additives. They are also concerned about the possibility of genetic modification (GM) contaminating organic products. The contentious issue of GM in organic food production is directly related to the preference for organic food among the European consumers (Torjusen et al., 2004).

Finally, it can be stated that consistent with more recent research findings, organic food consumption decisions are primarily driven by attributes such as health benefits freshness and taste (M. Wier et al., 2008). These researchers suggest that all of these attributes can be embodied through modern production, sales and distribution systems. However, it can also be argued that modern and contemporary production and marketing practices are common to all types of food and not only to the organic sector.

## Price sensitivity in organic food consumption

Organic foods are usually more expensive than conventional foods. On average, retail price premiums for certified organic foods range from 10 per cent to 15 per cent in Germany, 20 per cent to 30 per cent in Austria, 80 per cent in Australia and 10 per cent to 100 per cent in the US market (Lockie et al., 2006). The negative impacts of high prices of organic foods on organic food consumption are widely discussed in previous research (for example, Kumar et al., 2011). Not all researchers, however, agree that high prices negatively affect organic food consumption. Whilst some researchers suggest that consumers are willing to pay premium prices even for organic foods with less than 100 per cent organic ingredients (Batte et al., 2007), some others find that consumers are willing to pay up to 30 per cent more on organic foods than for their counterparts (Chinnici et al., 2002).

According to another group of researchers, willingness to pay premium prices for organic foods depends on purchase frequency (Ureña et al., 2008). These researchers suggest that whilst occasional organic consumers are willing to pay approximately 10 per cent higher prices, regular organic consumers are willing to pay approximately 15 per cent higher than for conventional food. Hence, it can be seen that willingness to pay relatively higher prices for organic food as compared to conventional food depends on some other moderationg factors such as purchase frequency.

Krystallis and Chryssohoidis (2005) further investigated the factors that influence consumers' willingness to pay for organic foods. These researchers report that whilst food quality, safety and trust in organic certification bodies influence consumers' willingness to purchase organic foods, prices and socio-demographic profiles of consumers have no relationship with their willingness to purchase organic food.

Researchers also suggest that consumers' willingness to pay for organic food varies depending on the type of product categories (Athanasios Krystallis and Chryssohoidis, 2005; Lin et al., 2008).

According to Krystallis and Chryssohoidis (2005) consumers' willingness to pay relatively high prices is lower for product categories such as tinned foods, pasta, yellow cheese, biscuits, feta cheese, bread, legumes, cured meat and milk. These researchers also report that consumers' willingness to pay more is higher for product categories such as fruits, vegetables, poultry and eggs. Another more recent study demonstrates that organic consumers in the U.S. are willing to pay from 20 per cent to 60 per cent more than conventional food, and this range of price differential varies in accordance with the product categories (Lin et al., 2008). Some very recent studies have confirmed this type of purchase behaviour which is associated with the product category of organic food (for example, Ngobo, 2011).

Ngobo (2011) also suggests that consumers are more likely to purchase organically produced store brands as compared to organic national brands and they are less likely to buy organic products that are often promoted in shop catalogues. The latter is consistent with previous research finding that willingness to pay for organic foods is highly correlated with consumption habits (Soler et al., 2002). In contrast, a more recent study reports that although purchase decisions of organic beef are associated with price and household expenditure, demand for conventional beef is associated with income, habits and seasonal consumption patterns (Anders and Moeser, 2008).

According to O' Donovan and McCarthy (2002), 70 per cent of organic consumers are not willing to pay more than 10 per cent higher than conventional food prices for organic foods. In a more recent study, Ngobo (2011) finds that the probability of purchasing an organic product is greater among the high income, college educated, and older families as well as among consumers with up-scaled occupations. Similar to previous research (for example, Anders and Moeser, 2008), Ngobo (2011) also suggests that larger families are less likely to buy organic products. This finding signals that high prices are still one of the significant barriers to organic food consumption. Ngobo (2011) further adds that although price plays a role in portraying the high quality of organic food, nevertheless it has its limitations in encouraging organic food consumption among the masses.

This section has discussed the price sensitive nature in organic food consumption. Consumers in different countries react differently to the higher prices and their willingness to pay more for organic food varies depending on the type of product categories. Also, owing to the higher prices, consumers of organic food may belong to the high income, college educated and older age demographic segments. However, price sensitive issues relating to organic food can possibly be resolved by improving the existing sales, distribution and marketing practices of the organic food sector.

## Influence of organic food labelling and certification on organic food consumption

Many researchers agree that organic labelling and certification play a significant role in influencing organic food consumption as trustworthy labelling and certification procedures guarantee the authenticity of organic foods (M. Wier et al., 2008). However, the existing labelling and certification procedures are apparently not very effective in reinforcing confidence in the minds of organic consumers. In fact, sometimes they rather confuse organic consumers. Due to various issues such as misinterpretation and improper labelling of organic food, and inconsistencies in organic certification procedures, consumers usually do not share a common understanding of the characteristics of organic food (Henryks and Pearson, 2010).

Organic food labels are one of the important sources of information from which consumers usually gain knowledge of the importance of organic farming (Bellows et al., 2008). This knowledge also drives organic food consumption. As such, some researchers find that given more accurate information about organic food through product labels, both consumers' trust in those labels and acceptability of labelled organic foods increases (Soler et al., 2002).

More recently, highlighting the misinterpretation of organic labelling, a group of researchers found various types of products that are commonly labelled as organic foods (Janssen et al., 2010). These include products with lower inputs of pesticides, food additives or concentrated animal feed. They are referred to as low-input products. These researchers analysed how consumers react to low-input products in a purchase simulation along with certified organic and conventional products (Janssen et al., 2010). It was determined that organic products were bought most frequently by more than half of the participants of this study in all product categories (vegetables, fruits, meat, eggs, milk, yogurt, bread and cereal, beverages), the low-input products (milk and yogurt) were preferred second and conventional products were the least preferred.

Using focus groups, Janssen and Hamm (2011) explored consumer awareness and perceptions of different organic certification bodies across five European countries (Czech Republic, Denmark, Germany, Italy and United Kingdom). It was found in general that consumers' knowledge of organic certification bodies was low across all the countries. More specifically, these researchers found that consumers in Italy and in the UK are unaware of the different organic certification bodies. Also some consumers in the Czech Republic, Denmark and Germany, showed preferences towards a particular organic certification body over others. Using a qualitative study, Henryks and Pearson (2010) report that there are discrepancies in Australian consumers' perceptions of organic food and their experiences, at various levels. For example, according to the study, many consumers believe that organic chicken is similar to free-range chicken and also that all products sold in farmers' markets are organic. Henryks and Pearson (2010) argue that the reason behind the aformentioned misconceptions are largely based on confusion over organic labels and their certification.

## The impact of lack of availability of organic food on consumption

Not having easy access to organic food is considered another significant barrier in organic food consumption. Not only do consumers face difficulties in seeking organic food with easy access, but they also do not have access to a vast variety of organic food to choose from when making purchases (Chinnici et al., 2002). O'Donovan and McCarthy (2002) found that if organic foods were available with easy access, 85 per cent of non-organic consumers would purchase them. Some other researchers also find that city dwellers are more likely to purchase organic foods than rural dwellers (Millock et al., 2004). It can therefore be stated that development of effective distribution systems is essential in encouraging organic food consumption (Kuhar and Juvancic, 2010).

Using the findings of sixteen in-depth interviews with consumers in Denmark, Hjelmar (2010) reports deep insights associated with the lack of availability of organic food. This researcher suggests that there are two forms of organic food consumption practices, i.e., convenience behaviour and reflective practices. It is found that convenience behaviour is common among pragmatic organic consumers who expect organic foods to be available in the local supermarket in a clearly visible place preferably with an eco-label. The pragmatic consumers expect minimal price differences between organic and conventional products. In contrast, reflective practices are common among politically or ethically minded consumers who are concerned about health, ethical matters such as animal welfare and environmental welfare. These consumers are also concerned about organic product quality and taste. Hjelmar (2010) also finds that reflective practices are motivated by life events and discouraging news associated with issues about conventional food.

## Lifestyles of Health and Sustainability (LOHAS)

LOHAS is an acronym for Lifestyles of Health and Sustainability, a market segment focused on health and fitness, the environment, personal development, sustainable living and social justice. A sizeable number of people belonging to this broad market segment have demonstrated a desire to consume organic food, hence it is appropriate to discuss here the characteristics and future trend of this important market segment. LOHAS builds on foundational work from the mid 1990's in the United States by sociologist Paul Ray. LOHAS describes a type of consumer that actively seeks out healthier and more sustainable lifestyle, product and service options and also the market for the products and services they buy. Ray called this group the 'Cultural Creatives' and described them as innovators and leaders of cultural change, voracious consumers of art and books and also major drivers of a type of consumption that demonstrated a recognition of individual and community impact (Everage, 2002). The values and worldview of people belonging to the LOHAS segment influences their lifestyle and this is depicted in Figure 5.1.

**Figure 5.1: Influences of LOHAS – Values, Worldview, Lifestyle.**

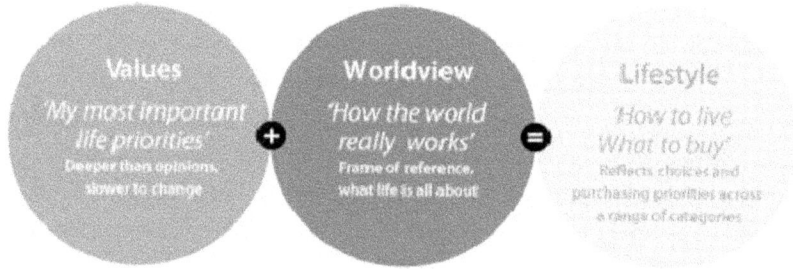

*Source: Paul H. Ray cited in www.mobium.co.au*

LOHAS consumers' lifestyle and purchasing decisions are informed by their values regarding personal, family and community health, environmental sustainability and social justice. However at the time of Ray's early work, there wasn't much recognition of the actual size and power of this segment as it was difficult to identify it using stereotype demographic profiles. In 1999, the founders of the LOHAS journal, i.e. Conscious Media coined the term LOHAS in an attempt to define the rapid emergence of a global trend of shifting consumer values based on sustainable, environmental and socially responsible platforms. Longitudinal research has revealed that these values, rather than being fringe and marginalised, are increasingly becoming broad based and entrenched across a wide cross-section of the population, driving changes in consumer markets, politics and society (Mobium Group's Living LOHAS Report, 2011).

Researchers have reported a range of sizes of the LOHAS market segment. For example, the Worldwatch Institute reported that in the U.S. this market segment was estimated at $ 300 billion, which was approximately 30 per cent of the U.S. consumer market (Worldwatch Institute, 2006). A study by the Natural Marketing Institute showed that in 2007, one in four adult Americans was part of the LOHAS group – nearly 41 million people. In the same year in Japan, approximately 17 million adults or 12 per cent of the population were LOHAS consumers (lohas-asia). Currently, in Australia the LOHAS consumer market is estimated at $ 21.5 billion and is growing rapidly. It is being driven by the consumption decisions of 4 million adult consumers and is impacting nearly every sector of the economy (Mobium Group's Living LOHAS Report, 2011).

**Findings of a recent Victorian study**

A recent study was conducted in Victoria, Australia (Lobo et al., 2011) to investigate consumers' purchase behaviour of organic food. Eight hundred and thirty nine respondents of this study reported the following:

> When asked what comes to their mind when they hear the term 'organic' food, a high percentage of respondents perceived it to be 'without chemicals' (67 per cent), followed by 'healthiness' and 'farming with nature' (10 per cent each), 'clean food' and 'other' (6 per cent each) and 'ethical food' (1 per cent). When asked what is their main reason for purchasing organic foods, respondents indicated 'health' (49 per cent), followed by 'local produce' (23 per

cent), 'other' (13 per cent), 'environment protection' (12 per cent) and 'ethical reason' (6 per cent).

When asked what types of organic food they have recently purchased, respondents indicated that they buy 'fruit and vegetable' (80 per cent), followed by 'meat' (35 per cent), 'tea/coffee' (20 per cent), 'grains' (18 per cent), 'milk' (16 per cent) and 'diary' (14 per cent). They were allowed to select more than one item, hence the aggregate of the percentages does not add up to 100.

When asked where they usually purchased organic foods, respondents indicated that they buy organic food from 'supermarket' (60 per cent), followed by 'farmers' market' (38 per cent), 'local stores' (23 per cent), 'organic food store', 'direct from farmers' and 'health food store' (12 per cent each) and 'fair food stores' and 'home delivery service' (3 per cent each). Again as they were allowed to select more than one item, the aggregate of the percentages does not add up to 100.

When asked about the frequency of their purchase of organic food, respondents indicated that they buy organic foods 'occasionally' (37 per cent), followed by 'once a fortnight' (18 per cent), 'once a week' and 'never' (15 per cent each) and 'more than once a week' (5 per cent).

Using the data of the same research study (Lobo et al., 2011) a Cluster Analysis (K-means) was used to segment respondents into distinct, but homogeneous groups, based on their perception ratings of the various attributes of organic food purchase. The three segments obtained were named: Pro-organic (32 per cent of the respondents), Reluctant (39 per cent of the respondents) and Organic Sceptics (29 per cent of the respondents) and discussion of their profile follows:

## Pro-organic consumers

Consumers in this segment strongly believed that organic food was good for their health. They held positive attitudes toward the nutritional value of organic food and also believed that they do not contain preservatives and chemicals. Hence according to them, organic food was ideal for children. In terms of price, these consumers seemed to be less concerned regarding the relatively high prices of organic food. Also, when compared to consumers in other segments, pro-organic consumers believed that organic food prices were fair and people should buy organic foods even though they cost more than conventional foods. In comparison to other consumer segments, they perceived that organic food tasted better than conventionally grown food and they were not put off by their unappealing looks.

Similarly, pro-organic consumers believed that if organic foods become more convenient to buy, they would buy more. They also strongly believed that organic food should become available in most food stores. They strongly believed that organic foods were good for protecting the environment and that more government support should be given to organic farming. They also moderately agreed that organic farming decreases soil degradation and that it is important to them to know whether the produce was grown organically or conventionally. In comparison with other consumer segments, pro-organic consumers strongly believed that it is important to support local farmers. They also believed in the food miles concept and only bought fresh food products when they were in season.In comparison with other consumer segments, pro-organic consumers supported the concept of fair trade. They would also be willing to pay higher prices to support small organic farmers.

As compared to other consumer segments, pro-organic consumers supported the concept of certification. They looked for certified organic logos when purchasing organic food. They moderately believed that organic certification programs have credibility. They had strong preferences for organic food produced in Australia and were hesitant to buy imported organic foods even though they might cost less. They also believed that only organic food products which cannot be produced in Australia should be imported. More importantly they needed to know where the imported organic products came from and the certification standards that they carried. They strongly believed that the government should do more to support organic farming and protect consumers from fraudulent organic claims. They believed that organic farming can replace conventional farming and that organic food is not a fad. They also perceived organic farming to be more ethical than conventional farming.

Finally, pro-organic consumers generally purchased organic fruits and vegetables, meats, grains and dairy. They purchased from all sorts of outlets including farmers' markets, organic food stores, local stores and health food stores. Their purchase frequencies were every week (28 per cent) and more than once a week (12 per cent). Their age demography was between 45-64 years (41 per cent of the segment). Finally they held postgraduate qualification (10 per cent of the segment) and secondary school education (27 per cent of the segment).

**Reluctant consumers**

Reluctant consumers had somewhat moderate attitudes towards health and nutritional values of organic food. Unlike pro-organic consumers, reluctant consumers and organic sceptics (segment 3) were price sensitive when it comes to their acceptance and purchase of organic food. They believed that organic food was too expensive and that they could not afford to pay more. However, as compared to the organic-sceptics, they believed that people should buy organic food (which cost more than conventional foods).

Reluctant consumers had some reservations about organic foods. They moderately believed that organic food tastes better than conventionally grown foods, but they were reasonably put off by their unappealing looks. Although they moderately believed that organically grown foods are better for the environment and that government should give more support to organic farmers, they personally did not care much whether such food products are grown organically or conventionally. In term of distribution (Convenience), reluctant consumers moderately believed that they would buy more organic foods if they become conveniently available in the food stores. They preferred to buy fresh food only when in season. Although they strongly supported the local farmers, they moderately believed in the concept of short food miles.

Reluctant consumers moderately supported concepts relating to fair trade and small organic farmers. Surprisingly, they did not believe that organic food which claims to be organic is all really organic and they perceived organic certification as lacking credibility. They also found that organic certification labels are confusing. Perhaps, these findings would suggest some attributes which hindered their acceptance of organic food products. With regards to their perception of imported foods, reluctant consumers had similar beliefs as pro-organic consumers (segment 1), except that reluctant consumers were more likely to buy imported organic products if they were cheaper, depending on the countries where organic foods are produced. With regards to their general attitudes towards organic food, reluctant consumers did not believe that organic farming could replace conventional farming. However, they believed that organic farming was more ethical than conventional farming. Although they strongly believed that the government should protect consumers from fraudulent organic claims, they only moderately believed that the government should do more to support organic farming in the future.

Generally, reluctant consumers perceived organic food as being 'without chemicals'. The main reason they purchased organic food was that it is locally produced. They also purchased more of organic fruits and vegetables. They tended to purchase organic products from many different outlets. They purchased organic foods occasionally (44 per cent of the segment) and were in the older age group (over 65 years old – 41 per cent of the segment).

## Organic Sceptics

Organic sceptics had somewhat negative attitudes towards organic foods. Compared with other consumer segments, organic sceptics had the lowest mean ratings with regards to health attributes of organic foods.

Similar to reluctant consumers (segment 2), organic sceptics believed that organic food was too expensive and they could not afford to pay more. Compared to other consumer segments, organic sceptics did not believe that organic food tastes better and they felt put off by their unappealing looks. With regards to distribution (Convenience), organic sceptics had the lowest ratings. They did not believe that they would buy more organic food if it was easily found in food outlets near their home/work. They also had a moderate opinion regarding the availability of organic food at most food stores.

In comparison with other consumer segments, organic sceptics had the lowest ratings regarding environmental protection attributes. They did not believe that organic farming protects the environment and decreases soil degradation. They gave the least support to the statement that government should give support to organic farming. They also indicated that it was not important whether the produce was grown organically or conventionally. With regards to regional products, organic sceptics were the least supporters of short food miles. They did not support local farmers nor did they purchase seasonal foods. Compared to other consumer segments, organic sceptics did not support the concept of fair trade and small organic farmers. With regards to certification, organic sceptics moderately believed that organic foods claim to be organic are really organic. They found organic labelling to be confusing. They also believed that organic certification programs lack credibility and thus, were unlikely to look for certified logo. They indicated that they were not aware of different organic food certification logos on the packaging. Interestingly, organic sceptics were more likely to buy imported organic foods if they were cheaper. They did not seem to mind where imported organic products came from and what logos they may have.

Organic sceptics had negative attitudes toward organic foods in general. They believed that organic food was a fad. Although they strongly agreed that government should protect consumers from fraudulent organic claims, they did not believe that the government should do more to support organic farming. Finally, organic sceptics associated organic foods with 'clean food'. Unlike pro-organic consumers whose main reasons for purchasing organic foods were health and farming with nature, organic sceptics purchased organic foods for other reasons such as taste. Their main purchase outlets were the supermarkets. A high percentage of organic sceptics indicated that they never bought organic foods (33 per cent). Organic sceptics tended to be younger and male (61 per cent of the segment).

## Conclusion

This chapter discussed specific factors which influence consumers' preference of organic food as compared to conventional food. These factors are health and safety concerns, price sensitivity, organic labelling and certification and lack of availability of organic food. Although there is no real medical evidence to suggest that consuming organic food is nutritionally superior to conventional food, nevertheless organic consumers tend to believe that organic food has more health benefits than conventional food. Hence they are willing to pay a price premium for such food. Some researchers suggest that consumers' willingness to pay for organic food varies depending on the type of product categories. In this respect, they are generally willing to pay higher prices for organic food categories such as fruits, vegetables, poultry and eggs. It has been ascertained by some studies that high prices are one of the significant barriers to organic food consumption. On the issue of organic food labelling and certification, it was generally noted that owing to misinterpretation, improper labelling and inconsistencies in organic certification procedures, consumers are confused and do not share a common understanding relating to the characteristics of organic food. Finally, not having easy access to organic food is considered another significant barrier in its consumption. Some studies have reported that if organic foods were available with easy access, a vast majority of the non-consumers would consider purchasing them.

This chapter then went on to discuss about consumers associated with Lifestyles of Health and Sustainability (LOHAS). These consumers' lifestyle and purchasing decisions are informed by their values regarding personal, family and community health, environmental sustainability and social justice. This is a growing market segment, especially in the developed countries which would have a significant impact on the consumption of organic food. This chapter finally concluded by reporting the findings of a recent Australian study, which dealt with several issues regarding organic food consumption, for example its usage pattern, reasons for its purchase, types and frequency of purchase. A cluster analysis of respondents in the Australian study revealed that three distinct segments of consumers existed. These segments were named Pro-organics, Reluctant and Organic Sceptics and the characteristics and profile of each of these segments was fleshed out. Overall the discussions in this chapter have important implications for marketers of organic food.

# 6
# Organic food supply chains: challenges and prospects

*Chandana Hewege*

## Introduction

The aim of this chapter is to explore the supply chain challenges faced by the organic food industry. As has been elaborated in the previous chapters, the organic food industry is fast growing due to the increased demand for organic products by the health-conscious consumers. Simply put, a supply chain facilitates the smooth flow of goods from the place of production to the place of consumption. Broadly speaking, a supply chain facilitates movement of goods from the supplier's supplier to the customer's customer. This means, a supply chain is responsible for the smooth flow of all the raw materials and components along the interconnected web of supply chain partners until the final product has been consumed by the consumer.

In order to ensure the efficiency and effectiveness of the organic food industry, it is crucial that the organic food is delivered to consumers at the right time, at the right cost, at the right quantity and at the right quality in the most efficient way. This can only be achieved through a well-lubricated and well-designed supply chain. Therefore, it is important to understand the specific needs of organic food industry and the related supply chain requirements. It is commonly accepted that the supply chain partners of organic food are faced with unique operational issues that may not be effectively addressed by conventional supply chains. Therefore, it is necessary to explore the unique requirements of organic food supply chain partners in order to help develop effective supply chains for the organic food industry.

The rest of the chapter is organised in five sections. The next section examines the importance of an efficient and effective supply chain with special reference to the organic food industry. Secondly, the chapter explores the current supply chain issues faced by the growers and processors in the industry. Thirdly, a detailed elaboration of the supply-chain related issues faced by the organic food consumers is presented. In the fourth section, the chapter outlines a way forward for the organic food supply chains.

## Importance of an efficient and an effective supply chain

> … we are pretty good at doing what we do, and that's growing organic products, and we really should not be spending too much time on the transport or marketing side of it (Excerpt from an interview with an organic farmer, Mascitelli, Lobo and Chen, 2011).

The organic farmer's comment above resonates with the perceptions of most of the organic farmers who tend to find it difficult to come to terms with the intricacies of managing the supply chain. Nevertheless, the crude reality is that the long term survival of an organic farming business is heavily dependent on the effectiveness of the supply chain which ensures making available the organic food to the final consumer at the right time, right quantity, right quality and right price.

The functions of a supply chain are not only limited to the physical movement of organic food from the place of production to the place of final consumption. Managing an organic food supply chain involves integrating the flows of products, information, and financials through the entire supply pipeline from the supplier's supplier to the customer's customer. An organic food supply chain can be viewed as a system of connected networks between the producers and the final consumer. In other words, it can be viewed as a series of integrated enterprises (eg. farmers, brokers, agents, wholesalers, retailers) that must share information and coordinate physical execution to ensure a smooth flow of goods, services, information and cash through the pipeline. In Australia, for example, the series of integrated enterprises consist of processors, wholesalers, retailers, export consolidators, agents, brokers, distributors and food service customers.

The logistics side of supply chain management expect to take care of the physical distribution of the organic food. Logistics management includes inbound logistics and outbound logistics. Generally, the logistics side of a typical organic supply chain performs a range of functions such as demand forecasting, purchasing, requirement planning, production planning, inventory control, warehousing, materials handling, industrial packaging, finished goods inventory, distribution planning, order processing, transportation and customer service. However, order processing, transportation, handling, storage and warehousing can be considered most critical functions, given the perishable nature of the organic produce.

The relative importance of logistics functions mentioned above is bound to vary depending on the scale of operations of the organic farming business. For example, a small-scale farm that caters to the local community may not face severe transportation issues, whereas medium-sized or a large organic farms need to focus more on securing an efficient mode of transport. Also, organic producers of fruit and vegetables, meat and milk have varied supply chain needs. For example, storage and transportation requirements of the dairy industry are significantly different from the requirements of the fruit and vegetable and meat industries. These issues will be explained in detail in the next section.

The most important reason for someone to be concerned about supply chain management is the supply chain costs that form a part of the total cost of the product and the total operational cost of a business. This means that supply chain performance directly influences the bottom line – *profit*. An inefficient supply chain not only adds to the total cost but also affects quality of the product and the customer service. The important criteria to judge the health of an organic supply chain are value and cost. The key questions that need to be posed are: does the supply chain add value to the product resulting in a higher level of customer satisfaction? or, does the supply chain operate efficiently effecting a reduction in the cost of the product?

Generally throughout the world, supply chains of organic products were used to be considered as alternative supply chains that are shorter and more locally based. Producers and consumers appear to be more tightly connected to each other than that of conventional food supply chains. Organic agri-food supply chains are networks of organisations that produce and sell fresh or processed products from vegetables, crops or animals. Organic food producers in Australia operate in a broad range of products that include dairy, fruit, vegetables, nuts, meat (beef, lamb, pork and poultry), wine, grapes and grain. With a view to ensuring a smooth flow of material, information and financial between supply chain partners, supply chains must be dynamic and flexible. Also, the supply chains must be built on cooperation, coordination, control, and trust. Organic food supply chains are more complex to design and manage than other supply chains (Ahumada and Villalobos, 2009a). Uncertain factors such as limited shelf life, perishability of products, risk of infestation, weather changes, strict quality and safety requirements and, demand and price variability tend to pose supply chain design challenges. If these uncertainties are properly managed, it will be easy to achieve supply chain coordination, competitiveness and customer service. When designing a supply chain, following factors should be considered (Stadtler, 2005);

- Selecting appropriate supply chain partners
- Identifying customer segments
- Determining the location of production and distribution facilities
- Identifying facility capacity and transportation means

Depending on the country-specific factors, organic food supply chain partners can consist of some or all of the suppliers of organic farming raw materials, organic farmers, brokers, agents, wholesalers, certification and regulatory bodies and retailers. When designing an organic food supply chain, one has to carefully select the appropriate supply chain partners. The number of partners, types of partners and their functional capabilities need to be assessed in terms of the supply chain objectives. Next, identifying customer segments is important because it defines the types of markets that need to be served by the supply chain. For example, an organic vegetable farmer may serve three customer segments; a restaurant chain, a retail chain and an organic food processing company. Each customer segment tends to demand different service levels. The organic supply chain should be so designed that it can cater to these three customer segments that have unique service requirements. Therefore, accurate identification of the needs of each customer segment is critical for achieving a superior supply chain design. The other important aspect of the supply chain design is the determination of the location of the production and distribution facilities. When dealing with fresh vegetables, fruits and meat, location of the farm, the major distribution mode and centres are crucial. Finally, it is equally important to ensure that the appropriate transportation facilities (e.g., cold storage facilities and trucks with temperature controls) are used.

Organic supply chain intermediaries, sometimes referred to as handlers, have a central role in the supply chain through their purchasing of ingredients, and packing, shipping, manufacturing, processing and distribution of organic products. These handlers or supply chain intermediaries work downstream with farmers and upstream with retailers. They are in a position to detect any problems related to supply of organic products and ingredients as soon as they occur. Since they have knowledge of the demand at the retail end, they can feed in valuable information to suppliers about specific market needs. These intermediaries tend to develop close relationships with organic farmers and other suppliers and as a result, can tackle demand-supply gaps effectively. Relationship among the supply chain partners is crucial here. Organic supply chain intermediaries add value to organic products as they move through the supply chain and also manage the additional requirement of maintaining organic integrity of the products (Dimitri and Oberholtzer, 2009).

**Supply chain issues faced by organic food growers and processors**

Organic food supply chain issues faced by the supply chain partners tend to vary from region to region, country to country or even industry to industry. However, there are some common issues that hinder smooth operation of organic supply chains throughout the world. In the USA, small organic growers have to constantly struggle to coexist with larger growers that have increasingly important scale economies (The Food Institute, 2005). Larger producers tend to sell their output to large food processors or retailers directly, or to wholesalers, whereas small scale organic food producers tend to use direct-to-consumer market channels, direct-to-grocery retailers and grower cooperatives. The organic suppliers lack marketing networks and sources of market information, making it difficult to obtain consistent supply. Product differentiation in the organic segment pose challenges with complex labelling laws and stocking fees (The Food Institute, 2005). A major difficulty in the USA is securing an adequate supply of organic products that are uniform in size and quality. Also, non-price factors are subject to asymmetric information during the negotiation process. An obvious solution to fill gaps between domestic supply and demand is to import organic products and ingredients, though international trade of organic food is hindered by the lack of synchronization of organic standards across the world (Dimitri and Oberholtzer, 2009).

According to Mascitelli et al. (2011), in Europe, the main problems associated with organic food supply chains are imbalance of demand and supply, high operating costs, lack of cooperation between partners and poor supply reliability. In Germany, Netherlands and France, it is sometimes difficult for consumers to locate and identify organic products, especially when majority of them are not sold in supermarkets. This is due to organic distributors' reluctance to cooperate with the conventional food distributors. However, in Denmark, Sweden and the UK, supermarket chains have partnerships with organic producers and encourage suppliers to produce and import organic products. In these countries, supermarkets play an important role in distributing organic products. Consumers in these markets tend to trust only products bought directly from farmers or regional producers.

Locally-based organic food supply chain partners face uncertainties such as poor collaboration, communication, and information sharing that cannot be reduced through the application of traditional supply chain design and management techniques (Kottila, Maijala, and Rönni, 2005; Tavella and Hjortsø, 2011). This is partly due to the fact that the traditional supply chain techniques do not sufficiently take into consideration some key aspects of local organic food supply chains such as ethics, sustainability and human values that affect decision making (Tavella and Hjortsø, 2011).

Local organic food supply chains consist of small-scale players that face limitations to implement complex mathematical models and sophisticated software used in traditional supply chain design and management (Ahumada and Villalobos, 2009b).

Difficulties faced by organic food supply chains can be viewed from four perspectives; relationship difficulties, communication difficulties, cooperation difficulties and economic difficulties. Table 6.1 shows these supply chain difficulties in detail.

### Table 6.1: Organic supply chain difficulties

| Perspective | Specific Difficulties |
|---|---|
| Relationship | 1. Difficulties in choosing right supply chain partners<br>2. Difficulties in finding supply chain partners with specific knowledge of organic food production and processing, management and economics |
| Communication | 1. Ineffective communication between supply chain partners<br>2. Lack of information sharing between supply chain partners<br>3. Difficulties in communicating the differences between organic and conventional products to end-consumer |
| Cooperation | 1. Lack of co-operation among supply chain partners<br>2. Barriers for small-scale enterprises to access supermarkets |
| Economic | 1. High operating, distribution and transportation costs due to small product quantities<br>2. Difficulties to allocate costs and returns to supply chain partners |

*Adapted from Tavella and Hjortsø, 2011*

Organic supply chain partners around the world often complain about lack of reliable supplies of organic raw materials, high transportation and distribution costs, difficulty in procuring sufficient quantities of organic products to distribute to retailers, difficulty in locating organic producers to buy from and obtaining shelf space in supermarkets (Dimitri and Oberholtzer, 2009). For small organic food producers, gaining access to major supermarket shelf space is a nightmare.

There is a growing trend around the world that large supermarkets are entering into organic food market creating both opportunities and challenges. The presence of large supermarkets is certain to increase the availability of organic food and the extent of land allocated for organic farming. However, the factors such as emergence of larger organic farms and firms, greater dependence on organic imports, pressure to lower the organic standards and the downward pressure on prices tend to adversely affect the organic farmers.

It is often the case that the information flow to consumers from the supply chain partners is weak. Supply chain partners share the view that it is the sole responsibility of the brand owner to communicate relevant information to consumers (Kottila et al., 2005). Also because of the poor information management and the information delivery to the consumers, the information relating to the environmental and ethical value of the organic product is not relayed among the supply chain partners (Kottila et al., 2005).

Like in any supply chain, organic food supply chains are also faced with the issue of non-value adding transaction costs. These are costs associated with supply chain activities that do not add value to the final product. The 'lean and mean' supply chain approach can deal with this issue and can reduce supply chain costs. However, when organic food consumers demand more convenience, diversity and competitive prices, organic supply chains need to turn to modes of vertical coordination or consolidation among the supply chain partners. The application of an integrated wholesale-retailer configuration with centralised buying operations is being considered as a viable option (Tondel and Woods, 2006). In the USA, the development of mass merchandisers and consolidation of retail chains have offered high buying power to a small number of large firms resulting in procurement issues such as retail 'price fixing' that lowers organic farmers' welfare (Tondel and Woods, 2006).

Supply chain consolidations can result in more direct buying from large suppliers who have access to automatic inventory management technologies. When retail partners consolidate, it signals wholesalers to consolidate or to enter into strategic alliances with a view to countering the rising buying power of the retailers. This would obviously benefit the organic supply chain in terms of achieving year-round supply, economies of scale and lower transaction costs. One of the emerging trends in the North American market is that organic food retailers tend to favour larger suppliers leading to a 'compressed supply chain' (Perosio, McLaughlin, and Cuellar., 2003). Direct buying from producers is an emerging trend.

In most parts of the world, organic food supply chains apparently face a lack of reliable supplies of organic raw materials and difficulty in procuring large quantities of organic products. While the world trends detailed above are broadly applicable to the Australian organic food industry, the nature and the intensity of some of the issues tend to vary. To elaborate further, inefficient storage, handling, transportation and distribution issues are prominent in the Australian context (Mascitelli et al., 2011). Moreover, there are obvious bottlenecks in the wholesale and retail sectors. These hindrances have prevented growers from achieving high sales volume and attracting future investments in this sector. Distribution of produce from 'farm to table' tends to have problems due to lack of efficient distribution channels from the main city to the point of purchase. For example, when a third party logistics (transportation) service is used, transporting the produce from farms to Melbourne is easy, but dispatching the produce from Melbourne to various outlets in Melbourne has several bottlenecks. This has motivated some growers to transport their produce on their own. Cooling facilities pose certain challenges as well. Growers do not have control over how their produce is handled and maintained until the produce reaches the final consumer. Particularly, there are issues surrounding the period from when the produce arrives at the wholesaler or retailer and the time they begin providing cooling facilities. The quality of the produce that the end-consumer experience is greatly dependent upon the rapidity with which the cooling facilities are made available to the produce. Most of the time growers or producers do not have control over this process. With a view to resolving these issues, it has been proposed that a dedicated distribution centre located either in Melbourne or in Sydney needs to be established (Mascitelli et al., 2011). This centre would be able to coordinate properly the distribution and storage related functions. On the other hand, many Australian organic growers complain that agents charge unrealistically high margins pushing the prices way above the level that can be afforded by the end-consumer.

Like in other Western countries, major super markets are negatively perceived by the Australian organic farmers who often complain that major supermarkets do not support or encourage the growers. The certification bodies that can play a significant role toward developing the organic food industry in Australia are criticised by many growers highlighting the fact that there are many farmers, wholesalers and retailers selling products as 'organic' where certification is not properly monitored. It is obvious that the industry will not grow as long as organic certification enforcement is not implemented. Majority of farmers are of the opinion that the commission or the levy that they have to pay for the certification is very high and does not justify the current level of benefits that they receive from certification bodies.

There are not many organic produce wholesalers operating in the Australian organic industry. For example, a recent study on Victorian organic farming industry reveals that the number of wholesalers operating in organic markets amounts to three companies whereas, in the conventional food industry there are over 700 wholesalers (Mascitelli et al., 2011). Temperature control is one of the major issues that wholesalers face when dealing with actual movement of organic produce from farm to the end-consumer. Apart from the cooling facilities, the cost of transport especially the cost of getting produce out of a farm onto a truck is alarmingly high.

**Organic Food Supply Chain Issues from the Consumer Perspective**

Triggered by a heightened concern over health issues caused by the use of chemical pesticides and other 'human unfriendly' processing methods, consumers around the world are increasingly seeking organic food. Modern-day consumers are sophisticated in their choices and preferences. They tend to demand a very high standard of service in terms of convenience, competitive prices, product features, availability, quantity and quality. These factors exert an enormous pressure on organic food supply chains.

From the organic consumers' point of view, two competing scenarios will have to be balanced. On the one hand, consumers need easy access to organic food. They need them at a competitive price and with a wide range of food varieties made available year round. It may be extremely difficult to achieve these service requirements unless the organic food supply chains transform somewhat into a conventional supply chain, which is essentially supermarket dominated. On the other hand, consumers would like to buy their organic food needs from local organic food vendor networks that may take several forms such as farmers' cooperatives, farmers markets and farm-gate sales. While the local organic food vendor networks ideally match with the 'green' and 'organic' perceptions of consumers, they may not necessarily be in a position to offer the much preferred service options that can be offered by the conventional supply chains. Hence the rise of large supermarkets selling organic food is seen as an emerging trend especially in the developed world.

It could be said that the 'supermarket' business model rightly suits the service needs of the sophisticated consumers who seek one-stop shopping facility enabled by wide range of products (often in excess of 30,000 products in one store), at relatively low prices with the added benefits of long opening hours and ease of parking (Pearson, Henryks, and Jones, 2011). Many supermarkets offer a wide variety of organic products ranging from private labels to branded organic labels. There has been an increase in the range of organic products sold in supermarkets (Pearson et al., 2011). This is certain to provide more choice not only for current consumers of organic food, but also non-organic food consumers may be lured to organic food market when they see organic food being made available widely.

Supermarkets' involvement in organic food supply chains is not a completely new phenomenon. For example, conventional UK supermarkets chains such as TESCO and Sainsbury were the pioneers in the sale of organic food in mid 1980s (Richter, 2004). In recent years, Sainsbury's involvement in organic food supply chains accounts for more than 65 per cent of all organic food sold in the UK offering over 1300 organic food lines (Sullivan, 2004). A similar trend is occurring in the European Union. For example, 75-90 per cent of organic food is sold through conventional supermarkets in Sweden and Switzerland.

Although some conventional supermarkets have captured a sizable share of national organic food sales, this is not the case across all supermarkets. According to a research finding, the level of commitment shown by conventional retailers (supermarkets) toward organic food sales tend to be shaped by the personalities and commitment of individual managers and related purchasing and marketing decision-makers (Kujala, 2005). These researchers further explain that the conventional supermarket chains' engagement in organic retailing is related to corporate identity and strategic plans, marketing orientation, national politics and the demographics of their customer base. Overall, the fast growth in organic food consumption is certain to have lured the 'profit-hungry' supermarkets into organic food retailing since the enormous growth in this market presents lucrative business opportunities- increased retail dominance by capturing and expanding the organic consumer market.

However, supermarkets are negatively perceived by many organic food consumers who prefer to buy their organic food from local speciality shops. In most situations, these outlets tend to emerge from community level interest in creating food supply chains that are resilient and at the same time supporting environmental, social and economic goals (Defra, 2008; Pearson et al., 2011). These 'local food outlet networks' often take the form of cooperative business structures, independent outlets, farm shops, farmers' markets, community supported agricultural schemes and productive gardens. These alternative organic food supply chains are environmentally friendly and farmer-friendly. Their organic food outlets tend to have short supply chains ensuring a higher return for organic farmers' efforts. Nevertheless, it is doubtful if the consumer would be benefited owing to inefficiencies and lower economies of scale.

When it comes to selecting an outlet to buy organic food, consumers tend to vary in their choice. At a glance, two types of organic food consumer groups can broadly be identified based on their organic food outlet preferences. The first group consists of those consumers who are anti-supermarket, pro-environment, farmer welfare concerned and resistant to large industrial scale operations. These consumers frequently patronise alternative organic food outlets. The second group consist of those consumers who seek convenience, wide variety, low price and availability. They are not averse to supermarkets. The current global trend, mostly in the developed world, suggests that the second group is becoming increasingly prominent. This undoubtedly supports the trend of the rise of conventional supply chains (supermarket-dominant).

New supply chain trends in organic food such as '100 mile diet' which means that people are encouraged to eat food from within 100 miles, will definitely create new challenges for organic food supply chains. This consumer movement aims to minimise the impact on environment, contribute to the local community and ensure high quality of organic food by retaining freshness. Organic food consumers who insist that the products they buy are sourced within 100 miles would favour those sellers who could verify the 100 mile claim. RFID technology can play a significant role in this regard.

In Australia, the 'food mile' concept is increasingly becoming a significant supply chain challenge. Australian organic food retailers are beginning to pay attention to the distance an organic food item has travelled to reach the retail outlet. For some products, 'food mileage' is printed on the label and the Australian consumers are beginning to be aware of the importance of this piece of information. The main advantages of the 'food miles' campaign to the organic supply chain is that the industry as a whole can benefit from stressing the importance of carbon dioxide reduction by lesser transportation and handling requirements, strengthening of local economies by protecting small farms, local jobs and local shops, and by increasing food security. However, there are serious counter arguments and counter-lobbying against the 'food miles' campaign. Some of the counter arguments are: high costs make local vegetable and fruits prohibitive for some; it ignores the environmental benefits of free trade; it jeopardises third world economies that rely on food exports; it is another form of protectionism. These counter arguments against the 'food miles' campaign which is lobbied by the non-organic industry stakeholders tend to pose severe challenges to the organic food industry. Yet, the organic industry can carve out a strategy to use the concept of 'food miles' to expand its customer base. For example, The Australian Conservation Foundation released a report entitled *'Food Mile Facts'*. This report establishes a strong case for 'food miles'(Australian Conservation Foundation, 2005 which is quoted in 'Food Miles' web as (UrburnFarmingOz, 2013):

> Measuring the full environmental impacts of food production, transportation, sale and consumption can be a complex task. However, several studies reveal some sobering facts about the hidden environmental costs of imported food.
> - The energy consumed in food freight often outweighs the nutritional energy in the food itself. For instance, it takes around 1,000 kilojoules of energy to ship 170kJ worth of strawberries from Chile to the United States.
> - A recent German study found that a 240ml cup of yoghurt in a supermarket shelf in Berlin entails over 9,000km of transportation.
> - In the United States, the food for a typical meal has travelled nearly 2,100km, but if that meal contains off-season fruits or vegetables the total distance is many times higher.
> - Even imported organic food can have a tremendous impact. A single Briton's shopping basket of 26 imported organic products could have travelled 241,000km and released as much $CO_2$ into the atmosphere as an average four bedroom household does through cooking meals over eight months.

Moreover, CERES – Centre for Education and Research in Environmental Strategies that maintains a community environment park in Melbourne, published a report entitled *Food Miles in Australia: A Preliminary study of Melbourne, Victoria* (Gaballa and Abraham, 2007). The report estimates the distances travelled for food items found in a typical Melbournian's shopping basket and the resulting greenhouse emissions from this transportation. The report explains:

> Food items like oranges, sausages, tea, baked beans etc with ingredients sourced from overseas have seen more of the world than most people. In fact, the report estimates that the total distance travelled by 29 of our most common food items is 70,803 km—that's nearly two times the distance around the Earth!

Despite the counter lobbying against Food Miles, it appears that the organic food industry can convince the consumers to buy the Food Mile argument. Consumer demand in 'organic processed food' creates a different supply chain route related to organic food – *the food service sector*. This sector caters to the 'eaten away from home' market. Included in this category are take-away food, food eaten in restaurants and pubs and institutional food that are eaten in schools, hospitals and other government organisations (Pearson et al., 2011). When complex food processing protocols are involved in the production of organic take-way food, consumers insist that the authenticity of organic ingredients is ensured. This responsibility rests on organic supply chains.

Organic food consumers especially in the developed world are increasingly concerned about the authenticity or 'truth' in the organic claims. Consumers would like to know who grew the organic food, how it was grown, what the environment looks like, how the animals and plants were groomed, how the products were packed and transported? These questions place an additional responsibility on the organic food supply chain to seamlessly communicate this information along the supply chain. For this to take place, organic food supply chains need to be fully transparent. Revealing supply chain partners' information to others would jeopardise the power that some supply chain partners exercise over other partners. Also, there could be confidentiality issues where supply chain partners could run the risk of losing their competitive advantage.

The importance of supply chain transparency was highlighted when a novel form of E Coli bacteria caused a serious outbreak of food-related illness in Germany in 2012. Scientific tests traced the illness to fresh vegetables contained in a food pack, but it was not possible to trace the source of the vegetables. Had the proper supply chain technology been used, tracing back to the supply sources would not have been a problem. Toward this end, some supply chain technology developers have introduced innovative systems where Radio Frequency Identification (RFID) is used to trace the organic product to its source.

## Way forward for organic supply chains

It has been highlighted in previous sections that the conventional supply chains are increasingly playing a dominant role in the organic food industry. Given the phenomenal demand growth in the organic food industry, organic food supply chains, whether it be conventional or non-conventional, need to be geared to satisfy the needs of the consumer. The organic food supply chains should not only be concerned with moving the goods but also geared to catering to the sophisticated consumer requirements. The ideal way forward for organic supply chains is to combine the best of both conventional and non-conventional (alternative) supply chains. In other words, the simplicity, authenticity, environmental friendliness, farmer-friendliness and local-orientation of alternative supply chains should be combined with convenience, wide variety, year round availability and competitive prices of conventional supply chains. This approach could cater to the modern-day organic food consumers' sophisticated needs.

A bright future holds for dedicated organic supermarket chains. Capitalising on the negative consumer perception toward conventional supermarket chains, there is a room for food chains solely dedicated to organic food to emerge as dominant players in the organic food market. These dedicated supermarkets need to combine the positive features of both conventional supermarkets and speciality stores.

As far as organic farmers are concerned, a multi-channel supply chain approach is certain to ensure their financial independence and bargaining power. This means that rather than supplying to one and only large supermarket or a buyer, organic farmers should have alternative channels such as farmers' cooperatives, farm-gate sales and internet (direct) sales while supplying to a large supermarket. By doing so, organic food consumers would also have alternative outlet networks to purchase their requirements without running the risk of being dominated by one or two supermarket giants.

Advanced supply chain technologies will have to be deployed with a view to ensuring quality control of the organic food. Radio Frequency Technology (RFID) will have to play a far greater role than what it is currently playing in the industry. RFID can also be used to ensure if organic food producers adhere to government food safety regulations. For example, in Australia, the National Livestock Identification System Legislation authorises RFID tagging for cattle stock. As a result of this technology being applied into organic beef industry, organic beef producers are now able to tag individual animals with RFID technology offering consumers peace of mind not having to worry about the authenticity of the supply chain and its sources.

Since a large proportion of organic food is now being sold through conventional supermarkets around the world with the support of global sourcing, organic farmers are required to adhere to strict international food safety regulations. For example, the European Union demands that all the organic food items should be capable of being traced throughout the supply chain from *farm gate to fork*. In the future, all the supply chain partners would need to reveal the identity of their suppliers. This could create some issues for some suppliers who derive competitive advantage by keeping their supply sources away from competitors and other supply chain partners in the same supply chain. Also, the supply chain partners should have mechanisms to withdraw and recall unsafe food items as soon as the food items have been identified as unsafe. One recent incident reiterates the importance of having a quick recall or withdrawal mechanism. In February 2013, a certain 'beef' product sold in a French supermarket had been identified as having some amount of 'horse meet'. A DNA test confirmed this. The supermarket has been ordered to trace the supply sources.

To avoid huge losses arising from product recall situations, the advanced track and trace facilities need to be introduced into the organic food supply chains. RFID technology can be used to automatically detect serial numbers or barcodes of batches of organic food that originate or are altered at distribution centres. Also, authenticity of the 'organic' claims can be verified by using this advanced technology.

According to a recent report entitled 'investigating supply chain management practices in the Victorian organic fresh fruit and vegetable sector', several gaps in the Australian Organic supply chain need to be addressed to ensure effectiveness of the supply chain. The meaning of 'organic' seems to be vague and blurred, especially from the customers' perspective (Mascitelli et al., 2011). There is a need for clearer government regulation and certification enforcement in this regard. In other words, the organic supply chain partners call for greater government involvement to ensure that all supply chain partners adhere to industry norms. There should be clearly spelt out rules to regulate the functions and obligations of the organic food certification bodies as well. The reason for this is that the organic food industry in Australia should clearly separate it from the non-organic (conventional) food. Currently, there are several competing certification bodies operating in the Australian organic food industry. The functions of these bodies should be coordinated in order to make their roles help bring the different partners of the supply chain together. Also, organic food supply chain in Australia is in need of a body or entity that creates stronger communication and network links between the separate components of the organic food supply chain. This is required in addition to the existing informal bodies and their informal networks representing growers, wholesalers and retailers. It has been highlighted that a new vegetable and fruit hub (as a collection point) needs to be established in the main urban centres in Australia. One of the essential features of these new vegetable and fruit hubs is temperature control facilities that ensure uniform quality of the produce from 'farm to table'.

## Conclusion

This chapter explained the supply chain challenges and prospects faced by the global organic food industry in general and the Australian organic industry in particular. It explained the essential features and functions of an organic food supply chain in general and elaborated organic food supply chain issues from both the supplier and the consumer perspectives. An organic food supply chain is a system of interconnected networks between the producers and the final consumer. It consists of a series of integrated enterprises that must share information and coordinate physical execution to ensure a smooth flow of goods, services, information and money along the supply chain. An efficient and effective management of organic food supply chain is bound to add value to both the supply chain partners and to the consumers.

From the growers' and processors' perspective, organic supply chain problems can take several forms. These problems tend to vary across countries and regions. It is challenging for small organic farmers to coexist with larger growers due to the disparity in the advantage of economies of scale. The organic food suppliers face issues such as lack of proper marketing network and sources of market information, inability to secure an adequate supply of organic products of uniform size and quality, imbalance of demand and supply, high operating costs, lack of cooperation between partners, poor supply reliability, poor collaboration and communication, and poor information sharing.

It could be said that the modern day organic food consumers are somewhat dualistic in their choice of the outlets. In an ideal world, they look for positives offered by large supermarkets such as convenience, competitive prices, wide variety, consistent shapes and sizes, and year round supply and also the positives offered by small locally-based speciality shops such as freshness, authenticity, environmentally friendliness, farmer-friendliness and personal service. This presents opportunities for supply chain innovation where all these benefits are offered to consumers through a single channel.

Driven by the enormous growth in the organic food industry, large supermarkets have already started to consolidate their dominance in organic food retailing especially in the developed countries. However, non-conventional supply chains which are locally-based coexist with large supermarkets. In order to cater to the growing consumer demand, organic food supply chains should be reinvented. Innovations could take the form of a hybrid approach where positive features of both supermarkets and small scale speciality outlets are made available to the consumers. Also, there is a room for chains of stores solely dedicated to handling organic food.

In the context of Australian organic food industry, there are several issues that need to be addressed in order to ensure an effective, reliable and cost-effective industry. Some of the salient issues are: distorted meaning of 'organic' in the minds of the consumers, need for a body or entity to create stronger communication and network lines between the supply chain partners, non-existence of fruit and vegetable collection hub in the main urban centres, proper procedures and regulated framework for organic certification and inadequate commitment from the stakeholders to promote the 'food mile' concept. The way forward for organic food supply chains will be to embrace sophisticated emerging supply chain technologies such as RFID. Supply chain transparency and tracking or 'tracing back to source' are going to be essential features of future organic food supply chains. These features will undoubtedly enhance customer satisfaction in terms of ensuring the authenticity of organic claims.

# 7

# Global trends in organics

*Andre Leu*

## Organic – the good news in the global downturn

Organic product sales, the number of hectares under production as well as the number of producers continues to increase despite the global economic slowdown. The information from most countries around the world is showing a consistent trend of a dynamic and growing industry. As with all trends there are examples of fluctuations in the rate of growth and even limited periods in some countries where there can be small declines, however the Meta data shows a strong active positive trend. Despite the current global economic downturn, the organic sector continues to grow and outperform most other agri-food sectors (Willer 2013). A snapshot of economic growth trend indicators are:

The global value of certified organic market sales was estimated to be US $62.9 billion globally in 2011. This is an increase from $59.1 billion in 2010 and US $54.9 billion in 2009. The comparison with the global organic sales of US $33.2 billion in 2005 and US $15.2 billion in 1999 shows a consistent trend of a high rate of growth (Willer 2013). Another important indicator is the number of countries that have formalised organic sectors whether for markets or production. 162 countries collected certified organic data in 2011 compared to 86 countries that collected certified organic data in 2000. The area of certified organic agricultural land (cropping, perennial horticulture and livestock, including dairy, poultry, pigs and gazing) was 37.2 million hectares in 2011. This compares with 36.2 million hectares in 2010, 25.7 million hectares in 2003 and 11 million hectares in 1999.

A significant emerging trend is the increase in land being used for organic wild collection. It allows wild areas to be certified for the sustainable harvest of wild collected products such as honey, mushrooms, herbs, spices, essential oils, berries, fibres and bamboo. It is also one of the few examples of a viable way to pay for the sustainable management of eco systems by ensuring a commercial return on the value of the eco system services and their products. This provides a strong incentive of a greater economic imperative to sustainably manage eco systems rather than clearing them for other economic uses or degrading them by over exploiting their capacity to sustainably regenerate wild products.

The current figures show that 43 million hectares were certified for wild collection in 2010. This is an increase from 41 million hectares in 2009, 26.5 million hectares in 2004 and 21 million hectares in 2001. The total area of land certified for the production and harvesting of organic products reached 80.2 million hectares in 2011. This includes agricultural land (37.2 million hectares, 2011) and wild collected areas (43 million hectares, 2010). There were 1.8 million certified organic producers in 2011 compared to 1.4 million in 2008 and 1.2 million in 2007. Over 90 per cent of the international market is in North America and Europe. However this is rapidly changing with emergence of markets in Asia, Middle East and Latin America. These regions are now the fastest growing markets for organic products in the world as the rising middle classes actively seek out pesticide free food due to food safety concerns.

The information for this snapshot of the data was published in a book titled *The World of Organic Agriculture Statistics and Emerging Trends 2013* by the Research Institute of Organic Agriculture (FiBL), Switzerland, in cooperation with the International Federation of Organic Agriculture Movements (IFOAM) (Willer 2013). The data for any given year (2011) are collected and collated the following year (2012) and published at the beginning next year (2013). The data are mostly restricted to the third party certified organic industry with a small amount of data coming from participatory guarantee systems. The true figures would be significantly greater than this as the majority of farmers using organic methods are not certified. An example are the consumer/farmer co-ops in East Asia that involve millions of consumers, several hundred thousand small holder farmers and have gross turnovers that would be worth several billion dollars. IFOAM and FiBL are investigating how to include this information to get a more comprehensive representation of the data and trends. However as there are very few official or published research statistics on the uncertified organic sector, it is difficult to generate credible data.

## Drivers of organic growth

The drivers of organic growth globally can be divided into two main categories: the pull from consumers and markets and the push from resilient production systems. Initially the growth in the organic movement was driven by farmers from the 1920s to the 1980s who were concerned about the loss of crop quality and economic viability due to the use of synthetic chemical fertilisers and pesticides. The resilience and appropriateness of organic systems, especially in terms of adaptation to climate extremes and food security continues to be a critical trend in the sustainable growth of the organic sector.

The publication of *Silent Spring* in 1962 raised the issue of toxic chemicals in food and in the environment. This was the beginning of the organic consumer movement due to their concerns over toxic chemicals in food. The consumer based market pull is a considerable driver of growth in the organic sector, especially with emergence of the third party certified sector in the 1980s. Organic guarantee systems are designed to ensure that consumers can trust the integrity of organic products and consequently have an important role driving demand. There are a range of new trends in these guarantee systems. Two other global trends are having an important influence on the growth of the sector. Despite the constant worldwide decline in the number of farmers there are steady increases in the numbers of organic producers due to their economic viability. Another important trend is the beginning of a science and research based approach after almost a century of being largely ignored.

## Part 1: The Pull from Consumers and Markets

The publication of Silent Spring in 1962 was the beginning of the organic consumer movement. The concern over pesticides is still the main consumer driver. Surveys by Newspoll, Nielsen and other credible organisations show that over 60 per cent of main grocery buyers in Australia and Canada, 78 per cent in the USA and substantial percentages in EU countries make some purchases of organic products, showing that there is a high level of recognition and acceptance of organic products (Newspoll 2008, Australian Organic Market Report 2010, OTA 2011).

This research shows that the increase in global demand for organic is being driven by consumers who are concerned about health, the environment and food quality. The primary reason for choosing organic food is due to health concerns with avoiding toxic chemicals as the main driver followed by the belief that organic food is more nutritious (OTA 2011).

This is part of the growing trend worldwide for healthier and safer foods. This rapidly growing market segment is called LOHAS (Lifestyles of Health and Sustainability). The organic sector has the largest dedicated and best known market share.

### Organic Guarantee Systems

The development of guarantee systems has been very important in growing this consumer demand. Consumers need to be able to trust the integrity of products labelled as organic. They are prepared to pay a premium for pesticide free food and need a guarantee on the veracity of the claim.

This led to the development of 3rd party certification systems where a recognised certification body checks that the producer is complying with claims. These systems started to become formalised and were introduced into government regulations in the 1980s. They have had a considerable role in driving the demand and credibility of organic products in the world market as they have built consumers' trust in the brand 'organic'.

## The Rise of the Developing World Middle Class

A substantial percentage of the rapidly growing middle class consumers in Asia, Latin America, Middle East, Eastern Europe and Africa are prepared to pay premiums for organic produce. Concern over food safety is the main driver for this increasing demand, especially concerns over unregulated pesticide use in most developing countries.

## Production for Markets

Many countries have policies that actively promote organic production as a way of accessing high value markets. Initially in many developing countries such a Uganda, Tanzania, Kenya, China, India, Peru and Colombia the emphasis was on developing high value export markets to Europe and North America. The growth in exports in these countries has seen the surplus organic production sold on the domestic markets. The advent of organic products on these markets has seen the development of domestic consumers' awareness of organic production and products. This has enabled consumers who are concerned about the food safety issues surrounding pesticides, to have access to products that are certified to a production system that prohibits synthetic pesticides. In some of these countries such as India and China, the growth of the domestic market due to food safety concerns has seen the domestic organic market overtake the export market in sales volumes and value. Officials from the Green Food Division of the Chinese Ministry of Agriculture stated at Biofach Shanghai in May 2012 and again at the Organic Trade Union of China Summit in Chengdu in November 2012, that the value of the Chinese domestic organic market was around US $8 billion compared to the export market with a value of around US $600 million (Wang Mau Hua 2012 Pers Com).

Bhutan is actively pursuing a policy of becoming a 100 per cent organic country because they see organic production with its prohibition on the use of toxic chemicals and synthetic fertilisers replaced with agro-ecological methods as fully aligned with their principles of Gross National Happiness. The government is actively assisting their farmers, the majority of people in Bhutan, to adopt good organic practices. The government is also developing domestic markets for these products to replace the conventional products that are imported from India. They see multiple benefits in this especially by ensuring pesticide free food and restoring the balance of trade in their favour. Bhutan currently runs trade deficits with India through importing a significant proportion of its food from there as well as importing a small amount of agri-chemicals to grow conventional food. Two of the significant benefits of organic agriculture will be the self-sufficient production of local food and the use of on farm and locally generated inputs rather that expensive imported inputs. The organic sector is rapidly developing in the Middle East and parts of Latin America, such as Brazil, with many governments actively supporting both the production and consumption of organic products.

## Broadening the Base of Organic to make it More Inclusive

The International Federation of Organic Agricultural Movements (IFOAM) is the global umbrella body for organic agriculture. It is composed of over 800 affiliated organizations in around 120 countries worldwide and represents 1.8 million third party certified organic farmers and substantially more uncertified organic farmers. IFOAM's goal is the worldwide adoption of ecologically, socially and economically sound systems that are based on the Definition and the Four Principles of Organic Agriculture.

## The Four Principles of Organic Agriculture

Organic agriculture is based on:

- The principle of health
- The principle of ecology
- The principle of fairness
- The principle of care

**IFOAM's Definition of Organic Agriculture**

The definition provided by IFOAM of organic agriculture is stated very clearly:

> Organic agriculture is a production system that sustains the health of soils, ecosystems and people. It relies on ecological processes, biodiversity and cycles adapted to local conditions, rather than the use of inputs with adverse effects. Organic agriculture combines tradition, innovation and science to benefit the shared environment and promote fair relationships and a good quality of life for all involved (IFOAM 2013. IFOAM includes all forms of agricultural systems that conform to the definition and the four principles as being organic. These include third party certification systems as well as other types of organic integrity guarantee systems such as Participatory Guarantee Systems (PGS) and Consumer Supported Agriculture (CSA) and other non-certified organic farming systems).

**Third Party Certification**

Third party certification systems involve producers being accredited to a standard or several standards by a recognised certification body that checks that the producer is complying with the standard/standards. Organic guarantee systems are important for consumers to ensure the integrity of the products that they buy. Consumers need to be assured that they are getting genuine organic products and that the claims are not fraudulent. Third party certification systems give these assurances and ensure the integrity of the organic value chains. Currently the majority of the $63 billion worth of products that has been documented as being sold in retail markets are third party certified. However their success has also come with a range of new issues that need to be addressed to ensure the constant growth of the organic sector.

**International Market Access**

The access to international organic markets is becoming increasingly difficult as more countries bring in organic import regulations. Currently 110 countries have organic regulations: 66 fully implemented; 19 finalized regulations but not yet fully implemented and 25 countries in the process of drafting regulations. The mature markets in Europe, USA, Canada and Japan that currently account for 90 per cent of international organic trade have organic import regulations. Emerging markets such as Argentina, China, India, Brazil, Mexico, Thailand, Philippines, Korea and Indonesia have or are about to develop organic import requirements. The numerous organic import requirements that are imposed on producers by governments are becoming obstacles to trade, which constrain organic market development. It means that exporters have to pay extra costs in both compliance time and fees to have specific certifications and inspections for each of these markets. The new regulations in Korea and China have significantly slowed market access to some of the fastest growing and highest value markets in the world. It is estimated that organic trade to China will decrease in the coming years due to introduction of stringent compliance requirements and significantly higher inspection charges that were implemented in July 2012.

Markets like the USA, China, Korea, Canada and Japan require that producers are directly accredited to their specific certification systems. These multiple requirements add considerable costs to the export of organic products and adversely affect competitiveness on these markets. They also make it difficult for smaller producers who do not have the economies of scale to access these markets as they cannot afford to pay for all the certifications needed for market access. There is a major concern internationally that the increase of organic regulatory import requirements are in effect non-tariff barriers that are severely affecting trade. The increasing burden of organic regulations not only restricts market access and dramatically increases the costs due to the need for multiple certifications It has the unintended effect of encouraging larger organisations with the economies of scale who can afford these extra costs at the expense of the smaller producers. This goes against the ethos of the original pioneers of the organic movement and is a key reason why IFOAM is actively working with the Global Organic Market Access project (GOMA) project to facilitate easier forms of market access.

## Global Organic Market Access project (GOMA)

The Global Organic Market Access project (GOMA) is a partnership comprising of the United Nations Conference on Trade and Development (UNCTAD), The United Nations Food and Agriculture Organization (FAO) and IFOAM. Its goal is to remove many of the barriers to organic trade by promoting the equivalence of standards and regulatory systems through the development and implementation of a range of equivalence tools.

**Figure 7.1: Global Organic Market Access Logo**

Global Organic Market Access

 IFOAM

*Source: IFOAM*

## COROS

IFOAM worked within GOMA to develop the Common Objectives and Requirements for Organic Standards (COROS). The COROS is a reference tool that allows the appropriate regional variations of standards while ensuring that these standards are equivalent to each other. The organic products produced in accordance with these standards are seen as equally reliable even though there are slight variations in their production requirements.

## Bilateral Equivalence Agreements

Several countries and markets are now doing bilateral equivalency agreements to facilitate market access and build the level of trade, rather than on insisting on direct compliance to their standard and certification criteria. Equivalence essentially means that products certified under the regulation of one country can be sold as organic in another country without having to be certified again to that country's specific standard. It removes the need for a second certification. The EU allowed a limited number of countries to have equivalency with its regulation when it was introduced in the late 1980s. Australia, New Zealand, Argentina, India and several other countries have equivalency agreements with the EU and with Switzerland. In February 2012 the USA and the EU signed an equivalency agreement that allowed certified producers on both sides of the Atlantic to have equal access to their organic markets. Canada has signed equivalency agreements with the USA and the EU and is in the process of negotiating them with several other countries. Many countries are currently negotiating equivalency agreements for the trade in organic products with their trading partners. Examples are the EU and China who are having discussions and Canada and India have also started the process.

## The IFOAM Family of Standards

IFOAM would like to see implementation of multilateral agreements as the next step from bilateral equivalency agreements so that one certification can access many markets. It has implemented the IFOAM Family of Standards using the COROS as a multilateral equivalence tool. IFOAM regards all standards that are assessed as equivalent to the

COROS as equivalent to each other. IFOAM is assessing both regulatory and private standards against the COROS to develop a list of equivalent standards to encourage the adoption of multilateral equivalence agreements. Saudi Arabia became the first country to accept the IFOAM Family of Standards in its regulation.

**Figure 7.2: The IFOAM Family of Standards**

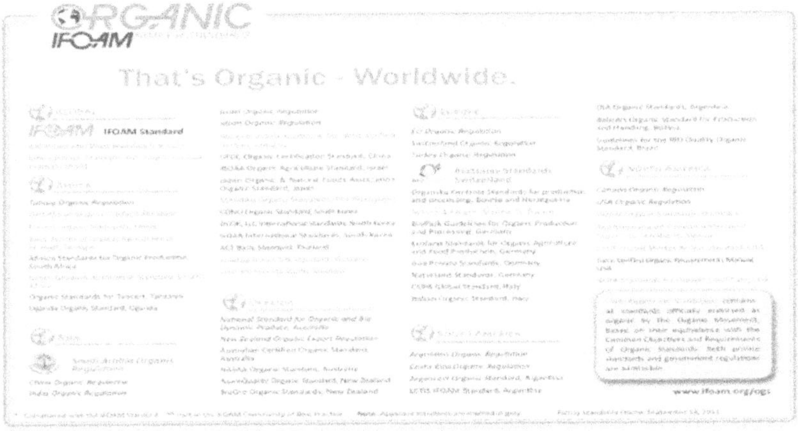

*Source: IFOAM*

## Regional Standards

A significant trend has been the development of regional standards that are adopted by the countries of a distinct geographic region. The advantage of countries within a region having one common standard is that it facilitates intra-regional trade by removing the need for organic producers having to comply with all the different national standards in that region. The first regional standard was the European Union organic regulation. All the member countries of the EU have to comply with this standard and consequently can trade within the EU without the need for multiple certifications to access markets in EU countries. ASEAN is currently looking at harmonising their organic regulations by adopting the Asian Regional Organic Standard (AROS) as part of the integration of their economies under the ASEAN Economic Community (AEC) in 2015. The AROS was developed through the guidance and support of the GOMA project.

The countries of Central America and the Caribbean have adopted a harmonised regional regulatory system for the trade in organic products. The island nations of the Pacific and counties of East Africa have developed regional standards and are currently in the process of developing regionally harmonised organic regulatory systems.

**Participatory Guarantee Systems (PGS)**

More than 90 per cent of the world's farmers are small holders on two hectares or less and live on less than US $400 p.a. In many cases the cost of 3rd party certification is higher than their annual incomes. This means that the majority of the world's farmers cannot utilise third party certification systems as a means of market access to achieve premium prices for their organic products. This is particularly important for the organic sector as the vast majority of organic producers are smallholders. One of the ways to lessen the costs is to 3rd party certify a group of farmers. This has worked well in many cases especially for exporting single products such as coffee or pineapples where the products of small holder farmers can be consolidated to achieve the volumes needed for the trade. However this was not always effective for local and regional markets where smaller amounts of a diverse range of products are produced and sold. IFOAM actively promotes Participatory Guarantee Systems (PGS) as a cost effective, transparent way of ensuring the integrity of organic products as a complementary organic guarantee system to 3rd party certification because they are suitable for small holder farmers. PGS systems are built around groups of producers who develop an internal control system to ensure that all the farmers in the group are adhering to accepted organic practices.

These normally involve a standard and a transparent peer review system to ensure that all the members follow the correct procedures. In many case consumers partner with the farmers to do the peer reviewed inspections.

**Figure 7.3: Examples of IFOAM Recognised PGS Logos**

*Source: IFOAM*

PGS systems are mostly built around local and regional markets, however two countries, Brazil and India have PGS regulations that allow the products to be sold nationally.

## Figure 7.4: Countries with PGS Systems

*Source: IFOAM*

## Other Non-Third Party Certified Guarantee Systems

The development of farmer and consumer based cooperative marketing systems are a significant global trend. The two best known examples of these systems are the Tei Kei system in Japan and Consumer Supported Agriculture (CSA) in Europe, North America and other countries. These systems work on a direct partnership between the consumers and the farmers. The integrity is built on the transparency and strength of the relationship between the two groups. The consumers are always welcome to visit the farms and in many cases there are open days where they can experience farm work. They are the shortest of all market chains and are built on the close relationship between the two parties.

Variations of these systems are the co-ops composed of both farmers and consumers. These can be done on any scale from a few farms and consumers to hundreds of farms with over a million consumers. The best examples of these are in East Asia, especially in Korea, Japan and Taiwan and involve millions of people in numerous co-ops. In Japan and Korea the bulk of organic sales are based on these systems and they are far greater than the sales of third party certification. They are worth billions of dollars however there has been no credible research to determine their true value.

An example of this is the Hansalim Coop in Korea. The coop has over 300,000 families as members which equates to over 1.2 million consumers with a gross turnover of around A$ 250 million. It also has around three thousand small holder famers as members. The prices are established every year through a meeting of the consumer and farmer representatives, with 85 per cent of the gross income going to the farmers and 15 per cent to run the organisation. Consequently the farmers have an income that is higher than the average Korean income and the consumers get fresh, local organic food at prices equal to or lower than the prices of conventional produce in supermarkets. This model proves that small holder farmers can be financially viable as well as highly productive. There are many co-ops of this scale in Korea, Japan and Taiwan that contribute billions of dollars to their national economies (Chang J Pers Com 2012).

**The Appropriate Choice**

The critical issue is for producers to be able to choose the system that is most appropriate to their circumstances. For many producers third party certification systems enable them to access high value regional, national and international systems. For others, particularly small holder farmers, the compliance costs can be higher than the financial returns so that PGS or consumer partnership systems are the most appropriate models. The key issue now is that farmers are able to choose the system that will bring them the greatest benefits. This is an important part ensuring that organic systems can be inclusive and available to all farmers.

## Part 2: Drivers of Organic Growth – Production for Farmer Viability, Resilience to Climate Extremes, Poverty Alleviation and Food Security

Some of the most significant drivers of growth trends in organic systems are in the areas of farm viability, resilience to climate extremes, poverty alleviation and achieving food security especially for smaller producers. The largest increases in the numbers of organic producers are in the developing world (Willer 2013). Several of the reasons for this are outlined below:

### Food Security

According to FAO and other figures the world produces more than double the amount of food to feed everyone. Despite this around 1 billion people suffer from hunger and another billion are malnourished, lacking the essential micronutrients they need to lead healthy lives. Clearly the current market based distribution systems are failing the poorest as they cannot afford to buy this food. The market based systems concentrate the food in the areas where people have the money to pay for it. Consequently one billion adults are overweight and almost half of them are obese (FAO 2011).

### Two Areas where Organic has High Yields

Research has shown two significant areas where organic systems have higher yields. These are under conditions of climate extremes and in traditional smallholder systems. Both of these areas are critical to achieving global food security.

### Greater Resilience in Adverse Conditions

Published studies show that organic farming systems can be more resilient to the increased frequency of weather extremes that are being caused by climate change and can produce higher yields than conventional farming systems in such conditions (Drinkwater, Wagoner and Sarrantonio 1998; Welsh 1999; Pimentel 2005). For instance, the Wisconsin Integrated

Cropping Systems Trials found that organic yields were higher in drought years and the same as conventional in normal weather years (Posner et al., 2008).

Similarly, the Rodale Farm Systems Trials (FST) showed that organic systems produced more corn than the conventional system in drought years. The average corn yields during the drought years were from 28 per cent to 34 per cent higher in the two organic systems. The yields were 6,938 and 7,235 kg per ha in the organic animal and the organic legume systems, respectively, compared with 5,333 kg per ha in the conventional system (Pimentel 2005). The researchers attributed the higher yields in the dry years to the ability of the soils on organic farms to better absorb rainfall. This is due to the higher levels of organic carbon in those soils, which makes them more friable and better able to store and capture rainwater which can then be used for crops (LaSalle and Hepperly 2008).

**Improved Efficiency of Water Use**

Research also shows that organic systems use water more efficiently due to better soil structure and higher levels of humus and other organic matter compounds (Lotter et al., 2003; Pimentel 2005). Lotter and colleagues collected data over 10 years during the Rodale FST. Their research showed that the organic manure system and organic legume system (LEG) treatments improve the soils' water-holding capacity infiltration rate and water capture efficiency. The LEG maize soils averaged 13 per cent higher water content than conventional system (CNV) soils at the same crop stage, and 7 per cent higher than CNV soils in soybean plots (Lotter et al. 2003). The more porous structure of organically treated soil allows rainwater to quickly penetrate the soil, resulting in less water loss from run-off and higher levels of water capture. This was particularly evident during the two days of torrential downpours from hurricane Floyd in September 1999, when the organic systems captured around double the water than the conventional systems captured (Lotter et al. 2003).

This is very significant information as the majority of the world's farming systems are rain fed. The world does not have the resources to irrigate all of the agricultural lands. Nor should such a project be started as damming the world's watercourses, pumping from all the underground aquifers and building millions of kilometres of channels would be an unprecedented environmental disaster. In many cases water from existing irrigation schemes is now being diverted for other uses such as environmental flows or for towns and cities. Improving the efficiency of rain fed agricultural systems through organic practices is the most efficient, cost effective, environmentally sustainable and practical solution to ensure reliable food production in the increasing weather extremes being caused by climate change.

**Small Holder Farmer Yields**

The other critical area where research is showing higher yields for good practice organic systems is in traditional small holder systems. This is very important information as over 90 per cent of the world's farmers fall into this category and the majority of organic farmers also come into this category. Two good examples are Uganda in Africa and India in Asia. Uganda currently has 188,625 organic farmers producing on 220,000 hectares which means the average size of an organic farm is 1.67 hectares. India has 547,591 organic farmers utilising 1.1 million hectares which mean the average size of an Indian organic farm is 2 hectares (Willer 2013).

A review of 114 projects in Africa covering 2 million hectares and 1.9 million farmers by the United National Conference on Trade and Development (UNCTAD) and the United Nations Environment Programme (UNEP) found that the average organic crop yield was 116 per cent higher than all African projects and 128 per cent higher than for the projects in East Africa (UNEP-UNCTAD 2008). The report notes that despite the introduction of conventional agriculture in Africa food production per person is 10 per cent lower now, than in the

1960s. Most significantly Supachai Panitchpakdi, the Secretary General of UNCTAD at the time and Achim Steiner, the current Executive Director of UNEP concluded that the evidence presented supported the argument that organic agriculture can be more conducive to food security in Africa than most conventional production systems, and that it is more likely to be sustainable in the long term (UNEP-UNCTAD 2008).

This is crucial information as the FAO data shows that 80 per cent of the food in the developing world comes from smallholder farmers such as those in Africa (FAO 2011). The developing world is also the region where most of the 1 billion undernourished people live, the majority of which are smallholder farmers. With a more than 100 per cent increase in food production in these traditional farming systems, organic agriculture provides the ideal solution to end hunger and ensure global food security.

## Tigray, Ethiopia

A good example of this is a project managed by the Institute of Sustainable Development in Tigray, Ethiopia. They worked in cooperation with the farmers to revegetate their landscape to restore the local ecology and hydrology. The biomass from this revegetation was then sustainably harvested to make compost and to feed biogas digesters. This was applied to the crop fields. The result after a few years was more than 100 per cent increases in yields, better water use efficiency and greater pest and disease resistance in the crops.

**Figure 7.5: Impact of using compost – Grain yields from over 900 samples from farmers' fields over 7 years**

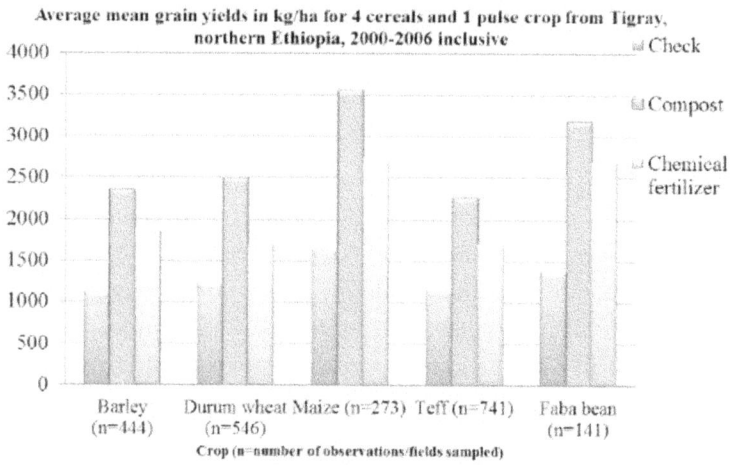

Source: Edwards et al. 2011

The farmers used the seeds of their own landraces which had been developed over the millennia to be locally adapted to the climate, soils and the major pests and diseases. These farmer bred varieties proved to be very responsive to producing high yields under organic conditions, whereas under conventional input practices they were susceptible to diseases such as rust. The major advantage of this system was that the seeds and the compost were sourced locally at no or little cost to the farmers whereas the seeds and synthetic chemical inputs in the conventional systems had to be purchased. The organic system had both higher yields and a much better net return to the farmers. This project using simple appropriate organic methods took a region that was previously regularly affected by severe famines that killed people through to a food surplus and relative prosperity. The people could now afford to eat well, to buy clothes, send their children to school, pay for medical treatment, afford transport into town and build adequate houses. The

Tigray Project started in 1996 in four local communities in the central, eastern and southern parts of the Tigray Regional State by the Institute for Sustainable Development (ISD). The Third World Network provided the initial funding. This project is still ongoing with ISD working with the Ethiopian Bureau of Agriculture and Rural Development and with Woreda experts and development agents to continue implementing the Tigray Project. The funding from several donor agencies is assisting the scaling up to enable hundreds of thousands of farming families in Ethiopia to adopt the practices (Edwards et al. 2011).

## Part 3: A New Science and Research Based Approach

A significant emerging trend is the increase in amount of science and research going into organic systems. Despite the fact the organic movement started at the beginning of the 20th century, these systems have been largely ignored by the scientific and research communities. Most of the production methods were developed by the farmers without any assistance from scientific research. Trillions of dollars have been spent on research into conventional agriculture while at the same time in the last 100 years there has been an almost total neglect of research into organic agriculture, agro-ecology and other farming systems that severely limit the need for pesticides. A significant proportion of this funding has been to develop and test the efficacy of synthetic toxic chemicals as pesticides such as herbicides, insecticides and fungicides.

Two comparison meta studies suggest that on average, organic yields are 80 per cent of conventional yields (Seufert et al. 2012, de Ponti et al. 2012). On the other hand a meta study by Badgley et al. suggests that the average organic yields are slightly below the chemical intensive yields in the developed world and higher than the conventional average in the developing world (Badgley et al. 2007).

Assuming that the analyses Seufert et al. and de Ponti et al. are correct, this is an incredibly small yield gap in relation to the enormous level of research and resources that have been spent to achieve it. The surprising fact is that millions of organic farmers have worked out how to get reasonable yields without the assistance of scientific research or the regular extension services that conventional agriculture receives. The main reason for the lower yields in some organic systems has been the fact that research and development into organic systems has been largely ignored. 52 billion US dollars is spent annually on agriculture research. Less than 0.4 per cent ($4 in every thousand) is spent on solutions specific for organic farming systems (Niggli U, Pers Com, 2013). Yet despite this lack of funding, the data sets from the meta comparison studies have examples of organic systems that have the same or higher yields than conventional agriculture.

**Figure 7.6: Frequency of relative yields of organic vs. conventional**

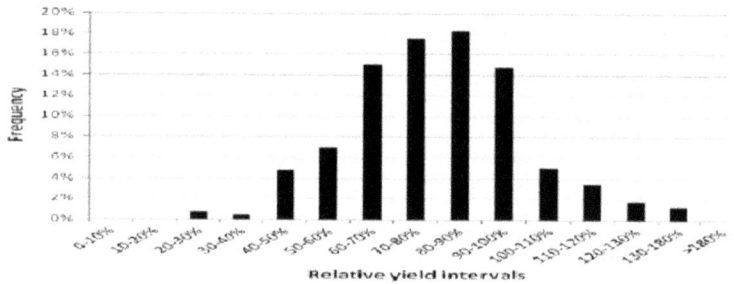

Fig. 1. Frequency of occurrence of relative yields of organic vs. conventional agriculture, grouped in 10% intervals.

*Source: de Ponti et al. 2012*

## The Urgent Need for More Research

Unfortunately in several countries the opposite is occurring. Instead of increasing the investment into organic research some countries are cutting back on it. Australia discontinued its meagre program in 2012 and the US congress reduced most of the funding in 2013 in its budget cut backs. Africa fortunately sees the multiple benefits of organic systems. The African Union Commission has adopted Ecological Organic Agriculture as part of the mix of solutions that are needed to achieve food security.

Given the small yield difference that has been achieved with trillions of dollars and countless thousands of researchers compared to what organic farmers have achieved largely with their own devices it would have to be argued that a substantial proportion of the funding into conventional agriculture has been a very poor use of valuable research funds. Also given that the new research into organic systems is starting to show very impressive increases in yields it is logical to argue that this is a far better use of these research funds. Fortunately the majority of countries in the world are increasing their levels of research and development into organic systems.

## Yields

One of the trends that is emerging from scientific research is that under the right conditions organic systems can have equal to greater yields than conventional systems. Some of the examples that have been published are below:

## Push-Pull System

The Push–Pull method in maize is an excellent example of a science based organic research that achieves substantial increases in yields. This is significant because maize is the key food staple for smallholder farmers in Africa, Latin America and in many parts of Asia. Corn stem borers are one of the most significant pests in maize. Conventional agriculture is reliant on a number of toxic synthetic pesticides to control these pests. Recently it has started to adopt genetically engineered varieties that produce their own pesticides. The Push-Pull system was developed by scientists in Kenya at the International Centre of Insect Physiology and Ecology (ICIPE), Rothamsted Research, UK and with the collaboration of other partners (Hassanali et al 2008, Khan and Picket 2010).

**Figure 7.7: Diagram of a Push Pull system in Maize**

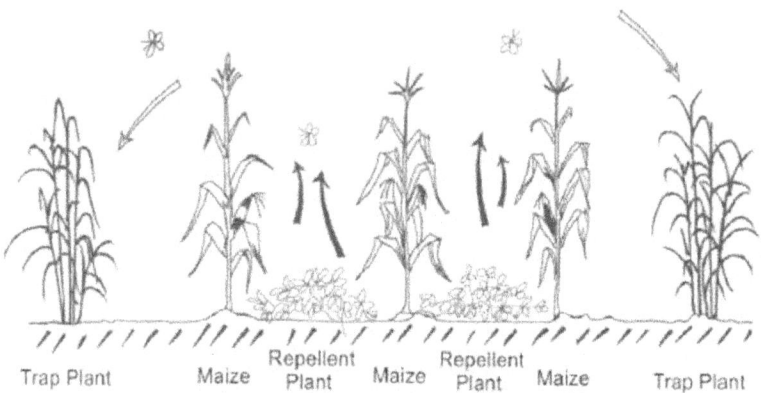

Trap Plant — Maize — Repellent Plant — Maize — Repellent Plant — Maize — Trap Plant

*Source: Khan and Picket 2010*

Silver Leaf Desmodium is planted in the crop to repel stem borer and also to attract the natural enemies of the pest. The Desmodium gives off phenolic compounds that repel the stem borer moth. Its root exudates also stop the growth of many weed species including Striga, which is a serious parasitic weed of maize. Napier grass, a host plant of the moth, is planted outside of the field as a trap crop for the stem borer. The Desmodium repels (pushes) the pests from the maize and the Napier grass attracts (pulls) the stem borers out of the field to lay their eggs on it in instead of the maize. The sharp silica hairs on the Napier grass also kill the stem borer larvae when they hatch, breaking the life cycle and reducing pest numbers.

### Figure 7.8: Silver Leaf Desmodium growing in Maize in a Push Pull System

*Source: Andre Leu*

Over 40,000 smallholder farmers in East Africa have adopted this farming system and have seen their maize yields increase from 1 tonne per hectare to 3.5 tonnes. This is more than a 300 per cent increase in yields and shows the huge benefits of shifting research away from toxic chemicals to science based ecological systems. High yields are not the only benefits. The system does not need synthetic nitrogen as Desmodium is a legume and fixes nitrogen. Soil erosion is prevented due to a permanent ground cover. Very significantly the system provides quality fodder for stock.

**Figure 7.9: Kenyan Farmer standing next to Napier grass that has been progressively cut and fed to a cow.**

*Source: Andre Leu*

One farmer innovation to improve this system has been to systematically strip harvest the border of Napier grass to use as fresh fodder for livestock. Livestock can also graze down the field after the maize is harvested. Many Push-Pull farmers integrate a dairy cow into the system and sell the milk that is surplus to their family needs to provide a regular source of income. Very significantly Push Pull has provided the farm families with food and income security and has taken them from hunger and desperate poverty into relative prosperity.

## US Agricultural Research Service (ARS) Pecan Trial

The ARS organically managed pecans out-yielded the conventionally managed, chemically fertilized Gebert orchard in each of the past five years. Yields on ARS' organic test site surpassed the Gebert commercial orchard by 18 pounds of pecan nuts per tree in 2005 and by 12 pounds per tree in 2007 (Bradford 2008).

## The Wisconsin Integrated Cropping Systems Trials

The Wisconsin Integrated Cropping Systems Trials found that organic yields were higher in drought years and the same as conventional in normal weather years. In years with wet weather in the spring the organic yields can suffer when mechanical weeding is delayed, and were found to be 10 per cent lower. This could be corrected by using steam or vinegar for weed control, rather than tillage. The researchers attributed the higher yields in dry years to the ability of soils on organic farms to more quickly take in rainfall. This is due to the higher levels of organic carbon, making the soils more friable and better able to store and capture rain (Posner et al. 2008).

## Scientific Review by Cornell University into 22 year-long Rodale Field Study

The scientific review found that:

- The improved soil allowed the organic land to generate yields equal to or greater than conventional crops after 5 years;
- The conventional crops collapsed during drought years;
- The organic crops fluctuated only slightly during drought years, due to greater water holding capacity in the enriched soil;
- The organic crops used 30 per cent less fossil energy inputs than the conventional crops (Pimentel et al. 2005).

## Rodale Organic Low/No Till

The Rodale Institute has been trailing a range of organic low tillage and no tillage systems. The 2006 trials resulted in organic yields of 160 bushels an acre (bu/ac) compared to the County average of 130 bu/ac (Rodale 2006).

## Iowa Trials

The results from the Long Term Agroecological Research (LTAR), a 12 year collaborative effort between producers and researchers led by Dr Kathleen Delate of Iowa State University shows that organic systems can have equal or higher yields than conventional systems. Consistent with several other studies, the data showed that while the organic systems had lower yields in the beginning, by year 4 they started to exceed the conventional crops. The researchers found that organic oat and alfalfa yields were 103 bu/acre and 4.4 tons/acre, respectively, compared to the county averages of 73 bu/acre and 3.3 tons/acre. Cost-wise, on average, the organic crops' revenue was twice that of conventional crops due to the savings from non-utilization of chemical fertilizers and pesticides (Delate 2013).

## Other Examples

Professor George Monbiot, in an article in the Guardian, 24 August 2000, wrote that wheat grown with manure has produced consistently higher yields for the past 150 years than wheat grown with chemical nutrients, in trials in the United Kingdom (Monbiot 2000). The study into apple production conducted by Washington State University compared the economic and environmental sustainability of conventional, organic and integrated growing systems in apple production and found similar yields (Reganold et al. 2001). In an article published in the peer review scientific journal, Nature, Laurie Drinkwater and colleagues from the Rodale Institute showed that organic farming had better environmental outcomes as well as similar yields of both products and profits when compared to conventional, intensive agriculture (Drinkwater 1998).

Dr. Rick Welsh, of the Henry A. Wallace Institute reviewed numerous academic publications comparing organic production with conventional production systems in the USA. The data showed that organic agriculture produced better yields than conventional agriculture in adverse weather events, such as droughts or higher than average rainfall (Welsh 1999).

Nicholas Parrott of Cardiff University, UK, authored a report, 'The Real Green Revolution'. He gives case studies that confirm the success of organic and agro-ecological farming techniques in the developing world (Parrott 2002).

- In Madhya Pradesh, India, average cotton yields on farms participating in the Maikaal Bio-Cotton Project are 20 per cent higher than on neighbouring conventional farms.
- In Madagascar, SRI (System of Rice Intensification) has increased yields from the usual 2-3 tons per hectare to yields of 6, 8 or 10 tons per hectare.
- In the highlands of Bolivia, the use of bonemeal and phosphate rock and intercropping with nitrogen fixing Lupin species have significantly contributed to increases in potato yields (Parrott 2002).

There are many examples of other innovative systems that are being developed in organic systems such as the System of Rice Intensification (SRI), organic no/low till systems (i.e. cover cropping and pasture cropping), agroforestry, holistic grazing, urban food production in sack containers and insectaries that are increasing yields. There is very good research that clearly shows organic agriculture can get the yields that are needed to feed the poor and the hungry. This is especially the case in smallholder agriculture – the majority of the world's farmers. The United Nations report concluded that all case studies showed increases in per hectare productivity of organic food crops and that this challenged the popular myth that organic agriculture cannot increase agricultural productivity (UNEP-UNCTAD 2008).

## The Technical Information Platform of IFOAM (TIPI)

In order to increase the level of research and development into organic systems IFOAM is working with the Research Institute of Organic Agriculture (FiBL), the International Society of Organic Agriculture Research (ISOFAR) and many other research organisations by forming the Technical Information Platform of IFOAM (TIPI). This global network of research institutions and universities will be able to share information and cooperate on projects to ensure the best use of the limited funds that are available for organic research. One of the key trends will be to get more farmers and scientists to work in partnership to improve existing systems and develop new ones.

## Part 4: The Increasing Number of Organic Producers

The numbers of farmers are declining around the world due to the lack of financial viability. This is particularly evident in developing countries where the mass exodus for rural areas is contributing to rapid growth of mega cities that cannot keep up with the necessary infrastructure. Organic agriculture is going against this trend of a decline in farmers and data shows a steady increase in organic producers. There were 1.8 million certified organic producers in 2011 compared to 1.4 million in 2008 and 1.2 million in 2007 (Willer 2013). One of the significant reasons for this is the financial viability of organic systems due a combination of resilient production systems and high value markets. Farmers need an income so that they can send their children to school, pay for medical bills, have adequate housing, clothing, transport and all the needs that we all aspire to. A viable income is an essential part of farm sustainability. Published studies comparing the income of organic farms with conventional farms have found that the net incomes are similar, with best practice organic systems having higher net incomes (Cacek 1986; Wynen 2006). The United Nations report found that organic production increased the amount of food production. It also gave farmers access to premium value markets and to use the additional income to pay for education, health care, adequate housing and relative prosperity (UNEP-UNCTAD 2008).

### Tigray

A good example is a cost benefit analysis conducted by the Institute of Sustainable Development of the conventional and organic farming systems in Tigray in Ethiopia.

Cost Benefit Analysis for the Farmer using Chemical Fertilizers:

- Costs in 2012 was US $300/ha for fertilizer (urea + DAP) and pesticides;
- Average yield of durum wheat grown with chemical fertilizer 4.5 t/ha;
- Sold at US $45/100 kg, farmers gross income would be US $2025;
- Net income after repaying credit US $1725 per hectare.

Cost Benefit Analysis for the Farmer from Using Compost:

- Average rate of compost application 80 sacks per ha (app 8 t/ha);
- Opportunity costs for making compost are virtually none as it is all family labour;
- Yield of durum wheat grown with compost 6.5 t/ha;
- Sold at US $45/100kg, farmers income would be US $2925 per hectare;
- All income stays with the farmer as there is no credit.

Other Benefits for the Farmer from Using Compost:

- Increased resistance to wind and water erosion;
- Farmers avoid debt from getting chemical fertilizers on credit – now costing USD 90 per 100 kg;
- Farmers making bioslurry compost can sell one sack (approx. 100 kg for USD 5.8);
- Competent farmers make over 35-100 tons a year (Edwards et al. 2011, Edwards Pers Com 2012).

## Philippines

A research project conducted in Masipag that compared the income between similar sized conventional and organic farms found that the average income for organic farms was 23,599 Pesos compared to 15,643 Pesos for the conventional farms (Bachman et al. 2009).

### Table 7.1: Improved Organic Productivity: Mean yield of rice, 2007 (kg/ha), n=840

|  | Masipag Organic | Masipag Conversion | Chemical Farming |
|---|---|---|---|
| Luzon | 3,743 | 3,436 | 3,851 |
| Visayas | 2,683 | 2,470 | 2,626 |
| Mindanao | 3,767 | 3,864 | 4,131 |
| Maximum | 8,710 | 10,400 | 8,070 |

*Source: Bachman et al. 2009*

### Table 7.2: Improved income from organic farm: Net agricultural income per hectare, 2007 (Pesos)

|  | Masipag Organic | Masipag Conversion | Chemical Farming |
|---|---|---|---|
| Luzon | 24,412 | 18,991 | 13,403 |
| Visayas | 22,868 | 16,039 | 13,738 |
| Mindanao | 23,715 | 17,362 | 19,588 |
| Average | 23,599 | 17,457 | 15,643 |

*Source: Bachman et al. 2009*

### Table 7.3: Improved income from organic farm: Annual Balance of Income and Expenditure per Household, 2007 (in Pesos), n=840

|  | Masipag Organic | Masipag Conversion | Chemical Farming |
|---|---|---|---|
| Luzon | 11,331 | 9,702 | -1,266 |
| Visayas | -1,090 | 287 | -4,974 |
| Mindanao | 5,481 | -232 | -7,546 |
| Mean Average | 5,967 | 3,407 | -4,546 |

*Source: Bachman et al. 2009*

While the yields are similar, the most significant information that came from this study was when the normal family living expenses were deducted from the net income. It showed that at the end of the year, on average, the organic rice farmers have a surplus income of 5,967 pesos whereas the conventional rice farmers had a loss of 4,546 pesos. This is one of the reasons why there is a continual exodus of conventional small holder farmers from the countryside to the cities and a continuing increase in the number of organic farmers. The data presented by the Institute of Sustainable Development in Ethiopia, Hansalim in Korea and Masipag in the Philippines shows that organic small holder farmers are viable and can enable these farmers to move from poverty to relative prosperity.

**Other Studies**

A study in the USA by Dr. Rick Welsh of the Wallace Institute has shown that organic farms can be more profitable than conventional farms. The premium paid for organic produce is not always a factor in this extra profitability. Welsh analysed a diverse set of academic studies comparing organic and conventional cropping systems. Among the data reviewed were six university studies that compared organic and conventional systems (Welsh 1999). The study into apple production conducted by Washington State University showed that the break-even point was nine years after planting for the organic system and 15 and 16 years respectively for conventional and integrated farming systems (Reganold et al. 2001).

**Conclusion**

The data on global trends shows a consistent trend of growth in the main categories such as the size of the market, the number of countries with organic sectors, the increases in the number of hectares used for production and the number of producers. This has been a consistent trend for around 100 years that has largely been driven by consumers and farmers with very little involvement from governments, scientists and the research communities. This started to change in the 1980's with the introduction of government regulations and formal research programs at universities and institutions. These two trends will continue to have a significant influence on the sector. The future will continue to see increased growth as the combination of government support, underpinned by good scientific research and driven by robust markets from strong consumer demand will see the organic sector move from being perceived as an irrelevant niche to a small and important sector. This will involve the multifunctional production benefits such as greater resilience, better water use efficiency along with good yields as well as a continuously growing market base of consumers who are concerned about the health and environmental aspects of food.

# 8

# A case study of trends in the Chinese organic food market

*Jue Chen and Barry O'Mahony*

**Introduction: an overview of the international market**

Since the mid-1990s organic food and beverages have grown from being an insignificant niche market to almost part of the mainstream. Organic products now account for an average of two per cent of food sales in Europe (though with countries like Austria, Switzerland and Denmark it reaches between five and seven per cent of retail value) and 2.5 per cent in the US. Developed markets currently lead the way with North America and Western Europe collectively making up 90 per cent of global organic sales in 2010. Within these regions, the US and Germany dominate organic sales, accounting for nearly 60 per cent of the total sales (Euromonitor 2011a). Organic products in emerging markets have remained limited, driven, in the main, by well-known staple products. According to "the world of organic agriculture: statistics and emerging trends 2012", global sales of organic food and beverages reached US$59.1 billion in 2010. Figure 8.1 shows that the global market has expanded almost three-fold between 2000 and 2010 and although growth slowed with the 2008 financial crisis, Willer and Kilcher (2012) have found that sales have continually increased at a relatively healthy pace.

**Figure 8.1: Global market for organic food and drink: Market growth 2000-2010**

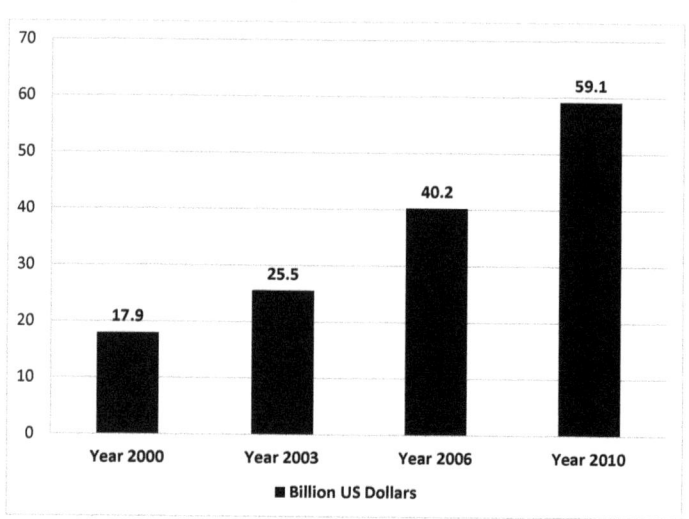

*Source: Adapted from Willer and Kilcher, 2012, p 122.*

In Latin America and the Asia Pacific region, organic beverages are also seeing growth mainly to meet the demand for hot drinks, specifically coffee and green tea. Dairy commands top position in terms of organic product sales, however, other products such as oils, fats, rice, baby food, sauces, dressings and condiments are also enjoying strong growth.

Previous studies have shown that consumers' concerns about health and the environment are among the most important factors that influence the purchase of organic food. Organic food products are perceived to be more nutritious, healthier, safer, environmentally friendly, free from harmful chemical residue and better tasting than conventional food. As a result, consumers are willing to pay a premium price for them even within emerging markets (Ureña, Bernabéu and Olmeda 2008). Perceptions related to health benefits are not well grounded, however, because there is lack of scientific evidence to support this claim (Kluger 2010). Indeed, research into organic food products has failed to provide any convincing scientific evidence demonstrating any health benefits over and above conventional food products (Brennan, Gallagher and McEachern 2003; Magkos, Arvaniti and Zampelas 2006). There are also a number of barriers to the purchase of organic food. These include consumers' perceptions of price (organic products are considered expensive), limited availability and choice, unsatisfactory appearance, misleading labelling and issues related to certification.

The purpose of this chapter is to provide insights into the market for organic food and beverage products in China and to report on current trends within this market. The chapter begins by outlining a brief history of the development of organic cultivation highlighting market size, regulations and standards. Further context is provided with reference to relevant literature and this is followed by a snapshot of the results of a large survey conducted in four Chinese cities which uncovers consumer attitudes and motivations in the purchase and consumption of organic food products with an emphasis on the most dominant factors. These were found to be logos and certification, price, lifestyle influences, ethnocentrism and food safety (Chen 2012).

## Introduction to the Chinese market context

The beginnings of organic food production in China can be traced back to the late 1980s and this was initially driven by environmental concerns within the domestic market. Growth in this agricultural niche was exponential with over one million hectares under cultivation within five years (International Trade Centre 2011). Concerns about food safety supported this growth which was reinforced by the introduction of government standards and regulations within the food sector. These early regulations paved the way for national standards on organic food production, which were introduced between the years 2000 and 2005. This created a framework to distinguish the organic food market from the mainstream agricultural sector, thereby providing additional confidence to buyers in relation to food safety (Chen 2012).

The principles of organic agriculture are similar to traditional sustainable farming methods which were a feature of early Chinese agriculture. However some of this expertise was lost as a result of the 'green revolution' in China. This began in the 1980s and was driven by a government push to achieve food security by introducing western farming techniques that were reliant on agrochemicals (International Trade Centre 2011). The vigour with which these farming methods were embraced led to degradation of soil, pollution of the water supply and harmful chemical residues on fresh food which entered the food chain causing major concern among consumers. Local government bodies sought to remedy these concerns by promoting eco-friendly farming techniques which paved the way for the China's organic agriculture industry (Xie and Xiao 2007).

The industry has been supported by a variety of government bodies since the 1980s which sought to protect the integrity of the organic food sector. State and privately owned agricultural enterprises invested in organic production when a number of European countries began to take an interest in Chinese produced organic products. Export opportunities emerged in Europe, the United States and Japan prompting Chinese government bodies to turn their attention to the regulation of organic food production (Xie and Xiao 2007). At the same time, the domestic food supply endured sporadic but significant food safety issues prompting the government to further protect the organic food market by introducing a number of food labelling schemes. These were designed to reassure purchasers of the quality of food products labelled under these schemes. Two types of labels were introduced: Pollution Free Food and Green Food. Pollution Free was designed to give buyers some confidence that food products with this label were compliant with general food safety standards. Food products labelled with the Green Food label were held to higher standards and certified by a public body. The Green Food standards were later expanded to include organic certification (Liu 2007).

While the expansion of the organic industry in China was facilitated by an export market, food safety concerns in the local food supply created a significant domestic market for organic produce. Although statistics relating to the export market are somewhat vague, between 2005 and 2008 significant rises in organic exports have been noted ranging from US$400 million in 2005 to US$800 million in 2006. In 2009, however, the export market declined due to the global financial crisis. This prompted organic producers in China to concentrate more on the emerging domestic market which has been growing rapidly since the mid-2000s (International Trade Centre 2011).

Indeed, the Chinese market is now considered to be a rising economic powerhouse and, along with a number of its near neighbours, has entered what has been dubbed 'the Asian Century'. These nations have experienced dramatic economic growth and are developing an affluent urban middle class which, amongst other things, is attracted to the purchase and consumption of organic food and beverages (Australian Government 2012). According to the Chinese Ministry of Agriculture, over the next decade China is expected to become the fourth largest consumer of organic food in the world. This will have a positive impact on several economies because a significant component of the current domestic organic market in China is serviced by imported products. These include organic wines, cosmetics and textile products equivalent to about US$500 million per annum. Despite the slowdown in many economies as a result of the global financial crisis, the Chinese market continues to show robust growth (Willer and Kilcher 2012).

The system of organic farming in China involves major conglomerates and corporations rather than individual farmers. As a result, it is more accurate to measure organic food production in China by the number of hectares under cultivation than by the number of farmers involved in the industry. However, there is almost no clear, comprehensive statistical data that specifies the true volume and value of organic production in China. Thus we are reliant on estimates which show that 2.03 million hectares of certified organic farmland was under cultivation in 2009. Figures for 2008 show that total organic production was valued at about US$2.4 billion of which US$500 million was exported and the rest was sold locally (International Trade Centre 2011). However it is difficult to find up to date and reliable information about organic production in China. Nevertheless, estimates provided by the International Trade Centre (2011) indicate that there was a major expansion of Chinese organic agricultural production in the mid-2000s rising from 101 thousand hectares in 2001 to 1853 thousand hectares in 2008. The most recent estimates suggest that somewhere between two and three million hectares of Chinese farmland is now certified as organic.

## Food safety as a driver of organic production

One reason for expansion of Chinese organic agricultural production in the mid-2000s is that Chinese consumers are losing confidence in locally produced food products as a result of a series of food safety breaches. In 2008, for example, melamine-tainted milk powder killed six infants and rendered 300,000 children unwell (US Department of Agriculture 2008). This food scandal involved one of China's largest dairy firms SanLu, and thus caused Chinese consumers to lose faith in the country's dairy industry. Another incident involving China's largest meat processor Shuanghui provoked the government to suspend its activities due to the presence of the illegal additive Clenbuterol[2] which was found in pork meat. The Clenbuterol pork meat scandal led consumers to express concern and disappointment in the meat processor giant. Previously both SanLu and Shuanghui were considered to be amongst the most reliable brands in China, but now Chinese consumers are unsure which

---

2 Clenbuterol is a chemical that can be used to prevent pork meat from accumulating fat. It is banned as an injected additive in pig feed in China as it can end up in the flesh of pigs and is poisonous to humans.

brand they can trust. Another incident occurred in December 2012 when China Central Television reported that some poultry farmers had fed chickens excessive amounts of antibiotics to help them survive in overcrowded coops. This was just another in a series of food safety scares that have eroded the confidence of Chinese consumers in recent years and have increased concerns about domestically produced food products (He, Yang and Shi 2013). This increased concern about food safety and caused this issue to become a top priority on the agenda of state councils. The dairy scandal also prompted well-off consumers to turn to imported alternatives. Thus concerns about trust in the food supply chain following these food safety scandals, coupled with the rising purchasing power of Chinese consumers has been a key factor in the growth of the organic food sector in China (Euromonitor International 2010b).

**Organic food usage in China**

The literature relevant to the consumption of organic food in China provides further context to this chapter and also provides additional perspectives in terms of interpreting the study results presented in the next section. This review found that there are five intended uses for organic food (Chen and Lobo 2012). These include for personal consumption, for children, for the elderly people, for gift giving and for family consumption. Personal consumption is listed as the least important attribute, while social status stands out the most important attribute for those that purchase organic food for the purpose of gift giving. If organic food is purchased for gift giving then the package, appearance and description of the product are considered important features. This is because gift giving in Chinese culture is an important element in the development of both business and personal relationships. The nature of the gift, in terms of price and prestige, can also represent the level of respect with which the recipient is held and signifies the financial status of the giver (Yau et. al 1999). In general, the high price of products implies prestige and is therefore intended as a sign of respect for the receiver, while simultaneously demonstrating the generosity of the giver (Sun and Collins 2006).

However, Chen and Lobo (2012) found that Chinese consumers could not identify significant differences in the appearance of organic, conventional or even genetically modified (GM) food products. As a result, marketers need to be aware that Chinese consumers will only understand and value organic food if improved, prestige packaging, that highlights the appropriate logos and the labels is used. Given the importance of gift giving among Chinese consumers in relation to reciprocity and the giving and receiving of favours, prestige packaging can be used to improve the receiver's understanding of the quality and value of the product. However, food safety is still considered as one of the most important attributes driving all five intended uses for organic products (Chen 2012).

## Legislations: Organic food standards in China

In China, the General Standard for the Labelling of Food (GB7718-94) (Chinese new food labelling law) has been in effect since February 1995 and closely follows the standards recommended by the Food and Agriculture Organization of the United Nations/World Health Organization's Food Law Committee. It requires that imported food labels are primarily in English or other foreign languages, and that the required content is presented in simplified Chinese characters. The Chinese national organic logo has been mandatory for all Chinese organic products since 2005, and this has been designed to make understanding organic labels more straightforward for consumers.

Another body, the China Green Food Development Center (CGFDC) was founded in 1992 to meet domestic consumer demand and to support the development of China's agricultural sector. The CGFDC is a specialised agency responsible for the development and management of green food under the supervision of the Ministry of Agriculture. This agency advocates: Green food products and includes edible produce and processed products produced in a sustainable environment. The CGFDC standards govern quality control, pollution, safety, and certification under two standards, 'A' and 'AA' (CGFDC 2012). 'A' represents a transitional level between conventional and organic food, allowing restricted use of chemical fertilisers and pesticides, while the 'AA' symbolises full organic status and matches all international organic food standards. A total of 98 per cent of green food produced in China belongs to standard 'A'. However, the 'AA' standard is not currently recognised by international markets as organic food.

The Organic Food Development Center (OFDC), founded in 1994, is the oldest and largest research, inspection and certification organisation in China. It is the only organic food certification body in China which is affiliated with the International Federation of Organic Agriculture Movements (IFOAM). Its mission is to safeguard human health and protect the ecological environment by promoting sustainable agriculture (Organic Food Development Center 2012). The Ministry of Agriculture oversees certification schemes in relation to green food, while organic food certification is jointly supervised by the Ministry and the State Environment Protection Agency. These agencies are fully supported by the Chinese Government. Indeed, in July 2002, premier Wen Jia Bao asserted that the government must regulate the certification and accreditation of non-polluted, green and organic food and accelerate support for modern production techniques (Gao et al. 2009).

In recent years, the Chinese government has also established a legislative framework in line with international organic institutions, with administrative rules and measures for organic product certification. The Certification and Accreditation Administration of the People's Republic of China (PRC) is a peak body set up by the State Council to perform the functions of administrative operation, unified management, supervision and comprehensive coordination of certifications and accreditations (Wang 2010).

## Chinese consumers understanding of organic food

The term 'organic food' has many different connotations and interpretations and in China the term is sometimes used interchangeably with the term 'green food'. Organic products are often referred to as 'eco-products', suitable for 'green' consumers who are ecologically or environmentally aware and concerned with health and quality of life issues. Generally speaking, in many western countries, 'green' connotes a sense of environmental sensitivity associated with global warming and the reduction of carbon emissions (Chen 2012). According to Shamdasani, Chon-Lin and Richmond (1993, p. 488). "The green consumer is generally defined as one who adopts environmentally-friendly behaviour and/or who purchases green products over standard alternatives". Other stereotypical images of organic consumers are 'greenies', 'health nuts' or 'yuppies', who are more interested in fashion than anything else. Consequently, consumption of organic food reflects a 'greening lifestyle' as well as a fashion statement for these consumers (Lockie et al. 2002).

Many developed countries have set stringent standards to ensure that organic food products conform to quality regulations. A number of developing countries—including—China have also established national organic product standards and regulations. In China three categories of food are deemed to be safe, as well as ecologically and environmentally friendly. These include 'non-polluted food' (or hazard-free food), 'green food' and 'organic food'. These three categories can be explained by means of a pyramid where the lowest level is non-polluted food, and the highest is organic food. The term 'organic food' has been influenced by the demands of the international market, while 'green food' and 'non-polluted food' labels have been introduced for domestic consumption. Figure 7.2 present examples of the labels that are used for the three categories (Gao et al. 2009).

## Figure 8.2 Labels of food categories in China

Non-Polluted Food
Green Food
Organic Food

This logo has only Chinese script.

This logo has both Chinese and English script. Most Chinese consumers are aware of this logo.

This logo has both Chinese and English script. Only a few Chinese consumers are aware of this logo.

*Source: Gao et al. 2009*

The percentage of organic food consumption (0.2 per cent) in China is considerably lower than the two per cent world average and organic food only accounts for about one per cent of total Chinese food consumption (Willer & Kilcher 2012). This is likely to be because food safety is a major driver among consumers and there are various 'safe' food categories which are recognised in China. As a result, 'organic food' is one safe alternative available within the domestic Chinese market. Most Chinese consumers are familiar with the concept of 'green' and 'non-polluted' food but many are not as familiar with the term 'organic food'. A survey by Li, Cheng and Ren (2005), for example, suggests that

92.2 per cent of respondents were aware of 'green food', 69.6 per cent knew about 'non-polluted food', but only 37.2 per cent were aware of 'organic food'. Organic food was also originally introduced to target the export market and, although China is the second biggest economy in the world, Gross Domestic Product (GDP) per capita is still only US$6,000 which is less than the world average (World Bank 2013). Consequently, the cost of organic food in the Chinese market may be a barrier because organic food is the most expensive of the three 'safe food' alternatives. Nevertheless, while only a small and seemingly insignificant percentage of the population consume organic food products, China has progressed from being a mere adopter of the global organic concept and is now an active organic food innovator (Gao et al. 2009 ).

## Survey of Chinese Consumers

The information provided in this section of the chapter is based on a survey which was conducted in 2010 in four selected first and second tier cities in China. The results of this survey are presented here to assist our understanding of the most dominant drivers of organic food purchase and consumption in China, as well as to highlight some barriers that prevent some Chinese consumers from purchasing organic food. The survey was conducted with urban Chinese respondents aged 18 and above who were familiar with the term 'organic food'. Survey participants were adults who were either regular, occasional or potential organic food consumers. These respondents were captured using intercept interviews in major supermarkets in Beijing, Shanghai, Shenzhen and Chengdu over a period of one month. A total of 964 valid questionnaires were obtained.

The four cities are geographically dispersed in the north, south, east and west of China and are economically and politically prominent because they are considered to be the main engines of China's phenomenal economic development. A profile of the survey respondents is shown in Table 8.1.

## Table 8.1: Profile of the respondents

| Demographic variable | Segments and Percentage |
|---|---|
| Gender | Female 60%, Male 40% |
| Age group | 18-30 years 49%, 31-45 years 35%, 46-60 years 12%, above 60 years 4% |
| Educational level | Postgraduate or above 10%, Bachelors' degree 40%, 2 years college or associate's degree 22%, High school 18%, Below high school 10% |
| Occupation | White collar 36%, Blue collar 14%, Not working 6%, Students 9%, Others 35% |
| Income | The monthly family income: 43% earned less than RMB 5000, 34% earned between RMB 5,001 and RMB10,000 16% earned between RMB 10,001 and RMB 20,000, 7% earned more than RMB 20,000. |
| Types of household | Single 21%, Couples without children 19%, Live with parents 24%, Live with young children 22%, Live with adult children 14% |
| Which city | Shanghai 29%, Beijing 24%, Chengdu 27%, Shenzhen 20% |

*Source: Chen, 2012*

Although the researchers sought to balance the sample demographic profile, the gender breakdown of these respondents includes more females than males. However, this can be explained by the fact that the survey was conducted with supermarket shoppers and, in China, women take more responsibility for family food purchasing and shop more frequently than men. The age group indicates that younger consumers prefer to shop in supermarkets and these are the major retail outlets for organic food products. The demographic mix of respondents also featured organic food consumers in urban China that are highly educated and high income earners.

**Can the logos and certifications be trusted?**

The results of the study show that Chinese consumers are strongly motivated by food safety issues related to personal health and are also concerned about the environment. They tend to be suspicious of the quality of most food including organic food certification because they don't believe they can rely on inspections and policy enforcement. However, they are eager to understand more about organic food and the organic food industry in general. While an organic food logo is usually the consumer's guarantee that the product has been produced organically, Chinese consumers doubt whether labelled organic products are really organic and whether they meet all declared standards. Chinese consumers believe that the organic market is not regulated by appropriate laws. These consumers demonstrate low levels of trust in the regulation of organic food and this may be because they have little understanding or awareness of organic food systems. They regard certification, quality, enforcement, information about nutritional value and food safety related to organic food as key issues when they consider their purchase.

Chinese consumers seeking information about organic food products during the pre-purchase phase seem to be interested in knowing where these products are produced. They seek to ascertain whether the organic logo can be trusted, and where they can buy available organic products. Organic buyers show more interest in tangible signs of quality which are reinforced by the label. This implies that regulatory attributes do significantly influence consumers' pre-purchase evaluation process.

This pre-purchase combination of assessing food quality through visual cues as well as examining logos and labels shows that some purchasers doubt the reliability of certification bodies and the agencies that are responsible for quality control. This is most likely to be influenced by the various aforementioned food scandals where deliberate acts have led to food contamination resulting in scepticism among about manufacturers' claims. Overall Chinese consumers are more suspicious of organic food products compared to consumers from many other countries. As a result, the Chinese government and food producers are being pressured to improve certification and inspection systems to regain consumers' confidence and trust. There is also a need for regulatory agencies to give clear and safe advice on food safety issues and for organic food certification bodies and marketers to ensure that information on food labels is clear and comprehensive.

The results of this survey also show that texture, appearance, smell, perceived nutritional value and fashion or trendiness are considered to be important attributes in the consumer decision-making process. This suggests that sensory appeal, for example, aroma and texture, are important to organic food purchasers. These, as well as non-tangible attributes such as trendiness, are part of the complex multitude of influences associated with the purchase and consumption of organic food. Perceptions of freshness was also found to be important to Chinese consumers and even more important is their mind-set that food is only 'fresh' when it has been purchased that day and has not been refrigerated. For this reason frequent shopping trips to purchase the freshest, tastiest and nutritious ingredients to make meals for family members are the custom in China. In summary, the freshness and texture are the quality indicators that have most influence on decision-making when purchasing food products and some Chinese consumers are also influenced by fashion and trends. As a result, they are reluctant to purchase products which have no visual signs of quality.

## Does price matter in China?

Previous studies suggest that the high cost of organic food is one of the barriers to its purchase and consumption (Chang & Zepeda 2005). The findings of our survey generally concur with Chang and Zepeda's study (2005), however, it was found that Chinese consumers do not have a great understanding of organic food. Despite this lack of knowledge they are willing to pay higher prices suggesting that, within the relatively affluent segment of urban Chinese society, price is less important than food safety and health related attributes. This further supports the argument that food safety is one of the most important issues influencing Chinese consumers' purchase of organic food products, at least in urban China.

Prices are elastic, however, and it needs to be understood that in China, prices of certified organic food and green food can be up to 700 per cent higher than those for conventional food products (Kluger 2010). Thus, while overall price sensitivity seems less important in terms of the purchase of organic food mainly because of food safety concerns, the results of this study indicate that retail price premiums are currently higher than the amount most consumers are comfortable paying. For example, more than half of the respondents (60.8 per cent) were willing to pay an extra 20 to 50 per cent, nearly 10 per cent were willing to pay an extra 51 to 100 per cent, but only one per cent of respondents were willing to pay an extra premium greater than 100 per cent. Approximately one quarter of respondents did not wish to pay extra at all for the organic products. This shows that most consumers' were prepared to pay premium prices of an extra 20 to 50 per cent over and above the price of conventional food products. It would, therefore, seem that organic food consumers in China have similar views about prices as those in Western Europe and North America where a 20 to 50 per cent premium was about the limit (Lockie, Halpin and Pearson 2006)

**Lifestyle influences**

Previous studies propose that lifestyle is a key influential factor in organic food consumption intentions. For example, Lockie et al. (2002) advise that there are stereotypical images of organic consumers and that they are often referred to as 'greenies', 'health nuts' or 'yuppies'. They also suggest that these consumers are more interested in fashion than anything else. As a result, consumption of organic food reflects a 'greening' lifestyle rather than a philosophic commitment. Our study investigated this dimension to determine whether organic food consumers in China might connect with this lifestyle segment. It was found that respondents to the survey do perceive organic food to be 'trendy', and many indicated that they are more interested in fashion than anything else. Indeed, one reason presented by these respondents for the purchase of organic food was its attraction as something new and 'fashionable'. This may suggest that the way organic food is marketed in the media reinforces the perception amongst some consumers that it is fashionable and trendy. These Chinese consumers' beliefs and attitudes were also fund to be influenced by two lifestyle segments: variety seeking and self-indulgence. Variety seekers are those that seek variety and novelty in their purchasing and consumption. They are keen to try something new and unique and enjoy a challenge. They are prepared to take a risk to experience new products and are strongly influenced by advertising (Yang 2004). Consumers in this segment are often the early adopters of new products and technology. The 'self-indulgence' category comprises those who make spontaneous decisions without deliberate thinking, and are those who do what and when they feel like (Lobo and Chen 2013). This category also considers that the main purpose of making money is to spend it.

The findings of this study suggest that organic food consumers in urban China are less traditional than their less urban counterparts and are looking for experimental experiences. They are risk takers who are individually centred and seek novel and new lifestyle experiences. Lifestyle issues are becoming an increasingly important dimension in China. The increase in the pace of home and working life in China has meant that health is a prominent driver, especially among office workers who are mindful of the need to balance work with fresh air, exercise and healthy lifestyles. As a result, healthy products and supplements are popular among middle-aged and young adults with relatively high incomes. These groups make up the elite in society, and are part of a big consumer group. There has been an increase in the number of people seeking a lifestyle of health and sustainability (LOHAS) which is a lifestyle trend emanating from the US. In China's large cities, consuming organic food, growing fruits and vegetables, and touring local farms has become popular. In Chinese this trend is called 'Nong Jia Le' which means 'vacation in the countryside'. The rising sense of 'enjoying life' and 'quality of life' is beginning to drive up spending, not only among consumers in key cities but also in the second and third tier cities and this places organic products in a good position to capitalise on this trend.

## Ethnocentrism and 'produced in China'

Ethnocentrism, which in this study refers to a preference for locally grown products, has been found to be another important motivator among consumers (Krystallis and Chryssohoidis 2005). Organic food is perceived to be environmentally friendly, which means that consumers may prefer to buy less travelled organic products from local sources rather than those that are imported (Fotopoulos and Krystallis 2002). Since this was likely to have a bearing on support for locally produced organic foods, a series of questions was included in the survey to determine if respondents in the study could be classified as 'Ethnocentric'. That is, whether these consumers prefer to purchase local and domestically produced organic food products rather than those that are imported.

The results show that Chinese consumers tended to show weak ethnocentric beliefs or attitudes towards organic food. The ethnocentrism construct per se does not appear to influence the beliefs and attitudes of urban Chinese consumers' towards organic food. These findings are supported by Hsu and Nien (2008) who suggest that ethnocentrism attitudes towards imported products in big cities in China tend to be diluted, particularly with those consumers who are more exposed to foreign products. Moreover, consumers from developing countries, such as China, sometimes tend to perceive domestic products as being of lower quality (Wang and Chen 2004). This is almost certain to be the case in China as a result of a lack of trust in locally produced food products. This is further supported by a previous market study which reported that in 2008, only 30 per cent of respondents trusted Chinese brands. This was down sharply from 44 per cent in 2007 (St-Maurice, Süssmuth-Dyckerhoff, & Tsai, 2008). In our study only 13 per cent expressed a clear preference for Chinese brands, indicating that in China's first and second tier cities, consumers tend to show no real affinity for locally produced organic food. It can be concluded therefore that although there are alternative organic equivalent products produced in China, the recent spate of food scandals has damaged consumer confidence in relation to domestically produced food products.

Our survey in four Chinese cities, revealed that nearly half of the respondents did not mind which country or region the organic food came from. However, if they were purchasing imported organic food, they indicated a preference for Australian and European products rather than Japanese or American. This might reflect in part Chinese consumers' animosity towards Japanese products and possibly some rivalry which exists with the United States. The results of the survey also reveal that Chinese consumers are more familiar with American certified logos than European and Australian ones.

Much has already been said about the various food safety scandals in China but it is worth reiterating here that issues relating to food safety seem to be paramount in the minds of consumers. This has prompted a number of agencies to review standards and the enforcement measures that support them and these issues have also been of interest within media circles. An investigation conducted within some organic supply chains in China by the Chinese Xinhua News Agency, for example, discovered that some food producers labelled products as organic without organic accreditation and that some unqualified manufacturers bought the authentication unlawfully (Pan, Wu, and Wei 2011). For this reason there is a lack of trust in the quality credentials of organic food and some doubt about the manner in which standards are enforced.

## Conclusion

Having outlined the context, previous research and current trends and issues in the Chinese organic food market a number of important elements stand out. Not least among these is the low percentage of organic food consumption (0.2 per cent) indicating that the market shows significant opportunity for growth. In order to capitalise on this growth, however, organic food producers need to understand the factors that influence consumers. These included certification, price, lifestyle influences and, most importantly, food safety, which was the main factor of influence for most respondents to the survey. Indeed, food safety has a major impact on almost all other issues, for example, brand or logo recognition, certification, which was generally considered open to manipulation, and price which was considered high but accepted in the interests of food safety. Some lifestyle factors were also seen to have an impact, however, these were more associated with novelty and fashion and, as a result, this might be considered a further niche market segment within this consumer group.

Dealing firstly with this issue of recognition of brands and logos, the food scandals in China have certainly raised awareness of the need for safe food. However, it was found that of the three product categories of safe foods ('non-polluted food', 'green food' and 'organic food') 'green food' was the most popular level of safe food and this may be related to price but it also reinforces the finding that as well as a lack of trust in the local certification regime, there is a lack of awareness of the international standards framework for organic certification. As a result, those suppliers that can overcome the lack of consumer trust in the quality credentials of organic food have the opportunity to position their organic products as leaders within this emerging market.

From a local perspective, the organic food industry needs to be aware of the necessity to continually monitor the organic food market and to ensure that strict standards are adhered to. In other words, they need to enhance the inspection and certification of organic food to ensure that labelling and logos actually guarantee quality. Organic producers and regulatory agencies should make a concerted effort towards improving awareness of the benefits of consuming organic food. Relevant stakeholders of this industry also need to implement market strategies aimed at enhancing consumer confidence and trust in organic labelling. Thus, there is a clear need to educate consumers regarding the differences between organic, conventional and genetically modified food in the market place.

From an international perspective, our survey found that ethnocentrism does not influence Chinese consumers' organic food purchase behaviour. While this may suggest a lack of trust in locally produced organic food, it should be noted that the Chinese Government introduced a new certification system in 2012 that allows the full traceability of the food supply and leaves the option open for possible verification by independent international agencies. While this will inevitably increase costs for producers, it is, nevertheless, an important step in building confidence for Chinese consumers and in building trust in organic products per se.

The lack of any emotional bond to products made in China among these consumers, however, does bode well for international organic producers. Currently, the main international players in the Chinese organic food market in China include the United States, Australia and a number of European Union countries. Our survey found, however, that there is a lack of awareness of what organic means, in some cases, and a lack of recognition of any real difference in the quality of international produce among these consumers. In particular, the study suggested that Australian and European brands were not well known indicating that these countries need to improve levels of recognition for their products and brands and, specifically, in the regulatory standards that are employed in the production of their organic products. Marketers, therefore, need to emphasise that their produce is safe, environmentally friendly and of high quality and this should be reflected in packaging, promotional material and branding. Differences between organic and conventional produce should also be highlighted.

Challenges within this market include perceptions of freshness, transportation and shelf life as well as some price sensitivity. However, respondents in this study did perceive that a higher price signified safe, clean products of high quality and, as a result, this price differentiator might be used to maintain organic products' image as a niche product provided that the premium remains within the 20 – 50 per cent range of conventional products. In summary, the Chinese organic food market is in an evolutionary phase and has grown exponentially since the early 2000s. The market presents major opportunities for both local and international businesses but consumer trust and brand awareness are significant issues that need to be overcome to ensure that the benefits of this market accrue to those that seek to access this marketplace.

# 9
# Contemporary issues related to food security

*Mark Gibson*

## Introduction

Over the past 40 years, since adequate records have been kept, the numbers of hungry and malnourished people around the world has stubbornly hovered between around 800 million and 1.2 billion people (SOFI 2011). Of this number many millions regularly die of hunger or hunger related diseases. Over recent years this has equated to around 10 million people annually of which nearly 6 million are children below the age of five. That equates to approximately one child's death every 6 seconds or so. By anybody's reckoning such numbers are indeed staggering and unacceptably high. Food security then – within this context and in a very general sense at this point – is about understanding just how this can still occur, especially when it does so under the very noses of an increasingly socially engaged public. Furthermore, the fact that this continues to be such a global problem to the extent that it does, particularly under the auspices of a vigilant global developmental community, is fast becoming one of the key challenges of our time.

So, one might reasonably ask: why the lack of progress? In answer, the following chapter aims to address some of the issues surrounding the difficulties in gaining a full understanding of the problems, some of the misconceptions and of course some of the different viewpoints which collectively serve to cloud rather than clarify one of the important social issues of the day. As a starting point much has been written regarding the lack of political will in effecting solutions; the complexity of the phenomenon; and the sheer scale of the problem among many other hindrances. In reality however, it is more likely the co-integration or interaction of one or more of these factors (Gibson 2012). Indeed one of the recurring complaints regarding the lack of progress surrounds this very complexity and in particular a general lack of understanding of the entirety of the issues involved. In short, despite the global reach of the phenomenon and the deceptively simple introductory definition, food security still engenders widespread misconception and misunderstanding; notably too among both professionals and interested lay parties alike (McCalla and Revoredo 2001, Gibson 2012).

One example of this confusion involves the simple use of the 'food security' terminology itself. This remains the case even in spite of much talk surrounding these issues within the media over the last few years (EuropeAid 2012). Instead it would seem that the term is still commonly being confused with similarly descriptive catchalls like 'food safety' and 'food sovereignty' to name just two (EuropeAid 2012). Another frequently perpetuated misconception by both the in-and un-informed is that food insecurity is the sole preserve of developing countries. Once again though, reality more closely resembles the contrary; in fact on this point it can be seen that many developed economies suffer the same inequalities of nutritional distribution as the more developing regions; although granted, largely to a lesser extent.

Apart from these examples though, available literature suggests there are more fundamental misconceptions at work here. The very complexity and cross-discipline nature of the problem has itself been charged with being the cause of much misunderstanding and the barrier to any real consensual solutions. Unfortunately the extent of this confusion is widespread and almost seems to be endemic as the following demonstrates:

> Malnutrition is poorly understood and has not benefited from expertise in communications and advocacy (WEF 2009);
>
> An analysis of the current literature on vulnerability make it apparent that there is ... no consensus on ... how to define and measure

vulnerability ... (Scaramozzino 2006);

... urban poverty and food insecurity – Urban livelihoods and coping strategies remain poorly understood (Floro and Swain 2010);

What are the water and food challenges faced by the world? Why are they so poorly understood? (Rijsberman 2010);

... food security and nutrition have not been given adequate attention [with regard to AIDS] despite broad agreement on their significance as an essential component of a comprehensive social protection package (Greenway 2008).

These are not isolated views either, a quick scan of the literature and one can readily identify many more instances of misunderstandings, subjectivity, misquotes and misrepresentation; and far too often sadly, without adequate recourse to fact (Budge and Slade 2009, Rijsberman 2010). Not surprisingly this fosters an environment beset with difficult analyses and divergent opinions; and not solely at the level of the individual either but many times at the institutional level too. Indeed one of the biggest failures in food security analysis over the last few years concerns one of omission on the part of USAID's Food for Peace (FFP) programme in 2006. By USAID's own admission they had, up until that point, failed to adequately take account of the notion that food security could be lost just as easily as it could be gained. After much self-reflection however, the notion of vulnerability was finally updated in 2008 and made explicit in their official food security definition (USAID 2007). While on the face of it such oversight in contemporary understanding almost defies belief, it soon becomes clear that the dynamic nature of the subject is such that this will probably not be the last admission, redefinition or reclassification of its kind.

## Multi-dimensional, Multifarious and Multi-Disciplined

Further compounding the problem of a full understanding of the phenomenon surrounds the cross-sectorial nature of food security itself. That is to say the concept is often defined or studied from a particular specialist angle; from single stakeholder lens' perspectives (Riely, Nancy Mock et al. 1999). Agriculturalists for instance might concentrate on tackling food insecurity through increasing food production by maximizing yield potentials. This might be achieved through improved crop varieties, through cross-breeding and hybridization or through disease and pest control measures etc. The agrarian perspective might also look to improve land management practices utilizing greater knowledge of agricultural needs and inputs. Economists on the other hand might consider food security in terms of opportunity lost. This might involve understanding the costs of malnutrition and lost labor to the economy or perhaps the cost of addressing the issues of food security more directly. Sociologists too might look for causality or symptoms of food insecurity among cultural and societal variables. These might include perspectives of population growth; how poverty affects malnourishment; the subjugation of women in developing economies; the psychology of income growth and changing dietary habits; or the dynamics of shifting rural-urban habitats; as well as the many non-food aspects of good nutrition in such areas of shelter, sanitation and education. Politicians too are involved in the issues of food security. Theirs however is perhaps amongst the trickiest of the perspectives as political considerations need to address multiple aspects of social and economic dimensions. As such their focus is usually more complex oftentimes aiming to implement policy that satisfies the greatest number of aims with any given limited resources.

This is not surprising as many issues surrounding food security wax and wane in importance and fall in and out of the public and political limelight. Therefore the political dimension to food security involves constant dynamic trade-offs between policies which hope to balance needs with objectives whilst reflecting the prevailing economic, social and political sentiments of the day; a challenging task. Then there are the theoreticians and academics who seek to understand, define and quantify the phenomenon. In doing so hypotheses and theories of causality and potential solutions are constantly promoted, altered and reworked as new evidence or theories emerge. Lastly, and certainly not exhaustively there are the numerous non-governmental organizations (NGO's) encompassing various charities and institutions whose focus is usually more frontline. Often with specific remits many seek to directly feed the hungry and malnourished or address underlying issues of poverty etc. This perspective also brings with it unique, complex and often dichotomous considerations of humanitarian aid.

The difficulty here is that while these separate and diverse perspectives, vis-à-vis issues of food security are all vitally important there is very little cross-sectorial research, especially when it comes to how these variables might interact at the holistic level. This is not surprising as one wouldn't for example expect an agriculturalist to bother himself with the daily per capita needs of the developing regions of South East Asia, nor would an economist be asked to study the optimum nutrient needs of transgenically crossed tropical-temperate plants. By the same token one would not expect a statistical modeler either to comprehend the subtleties of culturally acceptable food aid or a sociologist to ponder the food chain in regard to energy transfer up the trophic levels. So while researchers specializing in particular disciplines can and often do dip into other disciplines in the context of their research, food security analysis is such that a cursory dip, so to speak, would never be sufficient to properly comprehend its many facets. In this regard food security analysis and understanding tends to suffer (FAO 2000). While this might seem flippant or disingenuous it is not to denigrate the work of these vitally important disciplines, instead this simply serves to illustrate the often insular nature in which food security research is undertaken at the same time further highlighting the lack of an appropriate overarching perspective.

That said there are exceptions and it can be seen that the few who do take a holistic approach are institutions, think tanks and agencies such as the United Nations (UN), the Overseas Development Institute (ODI),

The International Food Policy Research Institute (IFPRI) and various humanitarian agencies or foundations like OXFAM, CARE, the Ford Foundation and the Red Cross for instance (Darcy and Hofmann 2003). However, even with many of these they tend to have priority areas focusing on one aspect or dimension of the bigger issue; and sometimes to the detriment of the whole.

With such widely differing perspectives then it is perhaps not surprising that opinions regarding such things as causality and solutions, come just as often from as many different directions as there are viewpoints (FAO 2002). For some, food security represents the ability to simply trade food in a global marketplace unimpeded by barriers of preferential trade agreements, import quotas or export tariffs. Others refer to food security as the right of country's food-sovereignty or its ability to directly or indirectly exercise control over its own food supplies; to determine what is produced and under what conditions. Yet others still, see hunger and nutrition issues as central to an individual's basic human rights and an area that needs strengthening. Whichever perspective one takes however, the salient point here is that food insecurity can be thought of as a multi-headed beast with many different masters (Earth Summit 2002, ActionAid 2003, FIAN 2010).

## The Challenge

The challenge and indeed the difficulty then lies in gaining an understanding of the full breadth and extent of the phenomenon from multiple perspectives from across these diverse disciplines. This is no easy task. Just take a look at the following table giving an indication of the kinds of considerations that food security practitioners must address. Looking at this table too it becomes a little easier to understand just how and why the confusion described above exists to the extent it does.

## Table 9.1: Issues of Food Security Consideration

| Conceptual | Understanding and modeling the phenomenon, misconceptions, increasing bundle of social aspirations, interaction and interplay, information dissemination. |
|---|---|
| Political and Economic | Globalization, poverty, governance and political will – (multilateral vs. unilateral, the right to protect), development paradigms (food production agricultural models), agricultural industry profile (intensification, extensification, small-scale farms, free trade, food sovereignty), rights and human capital development issues (right to food, rights of land, labor and women) considerations of natural and man-made disasters or emergencies – (war/conflict and social displacement, emergency response/aid, famine, displaced people), trade barriers (regulations/protectionism/nationalistic self-interest, lopsided or unequal trade balance), trading models (free trade, protectionism, food sovereignty etc.), a country's health (balance of payments etc., debt and structural adjustment), commodity/food prices, corporate control (transnational corporations), population growth, earth's carrying capacity, vulnerable groups (women and children, HIV/AIDs, refugees and displaced peoples), monitoring and early warning systems, food safety nets and food reserves, infrastructure (health and distribution of transport systems, markets, storage facilities etc.), food safety, urbanization: the rural–urban dynamic and humanitarian aid among many others. |
| Environment and Natural Resources | Water and land shortages, erosion and pollution, land use-changes – deforestation, sustainability, over fishing, food and biofuel competition, biodiversity, energy issues (sources, use, energy in agriculture), ecological accounting, land grabbing, climate change, natural resource footprints. |
| Science and Technological | Biotechnologies, agriculture, forestry and fisheries production technologies (research, development and dissemination), issues of intellectual property (biological patenting), green/evergreen technologies, genetic engineering (GM crops etc.), health and nutrition standards, monitoring and administration. |
| Socio-Cultural | Food and cultural relationship challenges, changing diets from wheat to meat, crop/ food wastage, environmentalism and sustainability, prevalence of undernourishment, care and feeding practices, coping strategies and priorities, education and employment, women: roles and gender parity and many more. |

*Source: The author, 2013*

Naturally from this, it can be seen that the notion of food security truly is a global phenomenon that clearly impacts on every human being's daily lives. In the opening paragraph a narrow and simple notion of food security in the context of the hungry and malnourished asked just how such numbers could exist. The proceeding paragraphs then highlighted the difficulties in answering this question. So, armed with a little more insight it is worth iterating the question: what is food security

## What is food security?

At the heart of the concept, stripping out all extraneous 'noise' so to speak, the basic premise of food security entails securing a regular supply of food to eat; this means not just for today or tomorrow but also for next month and next year too. Yet as hinted at in the preceding sections further delving into the nuances of the various aspects or perspectives of the phenomenon and it soon takes on wider, more involved meaning. A good place to begin to explore these ideas is through the lens of the Food and Agricultural Organization (FAO) of the United Nations. The FAO suggests that food security is the product of four interlinked constructs: food access; food availability; stability of supplies; and biological utilization (SOFI 2001).

In the first dimension – *food access*, the concept deals with notion that people have adequate access to their food requirements, both physical and economic and whether directly: through growing it; purchasing it; bartering or trading for it; or indirectly through other social arrangements such as through family; welfare systems; access to common resources; and finally through emergency food aid etc. (Sen 1981, USAID 2013). Ultimately, this idea can be thought of as forming a package of food entitlements that allow individuals, families or groups to acquire and maintain appropriate foods for an adequate and nutritional diet. In short, access refers to the ability of households or individuals to purchase or produce sufficient food for their and their families own needs.

When it comes to the dimension of *availability of food*, it can be seen that no matter the country or location, all food is provided through only one of two means – domestic production or imports (or a combination of both). As such, thought and consideration must be given to the physical availability of food at farms and in local markets. Movement of such foods too must be facilitated through well-functioning infrastructures of market, road and rail as well as ensuring there is sufficient and adequate storage and processing technologies throughout the various systems (Riely, Mock et al. 1999, FANTA2 2010).

Although not a new idea, when it comes to *stability of food* supplies, the realization that food security can be lost as well as gained is of increasing concern (USAID 2007). This brings into the fold the notion of risk management as a tool in the fight against hunger and malnutrition. In turn considerations of stability and vulnerability need to be made vis-à-vis the wider political environment and economy in general; of livelihoods and incomes; and even of food supplies in situations of emergencies such as floods, droughts or pests etc.

The last concept – *biological utilization*, concerns the ability of a person to properly absorb, at the biological level, the food they eat. This ability has been shown through research to be closely related to a person's health status. In turn it is understood that this health status is not solely based on the food an individual eats but also on the important non-food inputs as well. In this way any person's optimum biological utilization of food is predicated on adequate knowledge of nutritional and physiological needs as well as the proper application of such knowledge (FAO 2006, FEWSNET 2010). This necessitates considerations not only of food access, supply, and stability but also on considerations of health and child care; clean water and sanitation services and a suitable infrastructure to administer such services.

Collectively these four dimensions form the core of many private and institutional definitions describing the food security phenomenon. A cursory look at the various definitions serves to highlight the extent of conceptual similarities and differences held by these various bodies. Although before that, for the sake of perspective it is worth noting that even back in 1992 a thorough study by Maxwell and Frankenberger had already identified close to 200 separate definitions of food security (Maxwell and Frankenberger 1992). However, staying with the more widely accepted institutional classifications on offer, there are perhaps two major bodies whose definitions are amongst those most commonly quoted; those of the United Nations (UN) and various bodies of the United States.

Firstly, the UN's definition has had many incarnations over the years with the last update being refined in The State of Food Insecurity (SOFI) 2001. In this annual publication it was suggested that:

> Food security [is] a situation that exists when all people, at all times, have physical, social and economic access to sufficient, safe and nutritious food that meets their dietary needs and food preferences for an active and healthy life (FAO 2003).

In the second, the US take advantage of several definitions each worded slightly differently reflecting different institutional bodies' needs. In general though two main bodies: the US Department of Agriculture (USDA) and the US Agency for International Development (USAID) are the predominant stakeholders in this field. While both deal with issues of food security, the USDA usually focuses on national hunger issues while the USAID mainly operates at the international level.

In the USDA definition, food security is defined as:

> Access by all people at all times to enough food for an active, healthy life. Food security includes at a minimum: (1) the ready availability of nutritionally adequate and safe foods, and (2) an assured ability to acquire acceptable foods in socially acceptable ways (e.g., without resorting to emergency food supplies, scavenging, stealing, or other coping strategies)" (Andersen 1990, USDA 2009).

Another of the USDA's arms, the Economic Research Service (ERS) further define the food insecure as those consuming less than 2,100 kcal per day (USDA/ERS 2010).

Similarly, the USAID also employs several definitions dependent on need and purpose. However, generally speaking, USAID's current all-purpose classification is based on USAID Policy Determination #19 from 1992 which suggests food security exists:

> When all people at all times have both physical and economic access to sufficient food to meet their dietary needs in order to lead a healthy and productive life" (USAID 2010).

The US also promotes food security through the Agricultural Trade Development and Assistance Act, or more commonly the Public Law 480 (PL 480) programme. This incarnation offers a more flexible definition thus allowing for a range of possible interventions. Once again this definition is also based on Policy Determination #19 defining food security as:

> Access by all people at all times to sufficient food and nutrition for a healthy and productive life (USAID 1992).

There are of course others including those of the European Union, the various NGO's and the like. However, these definitions were chosen for their popularity among food security practitioners and the fact that they collectively represent the more convergent of the many definitions on offer (Gibson 2012). Yet, even in spite of such seemingly similar conceptual components there are also in fact many key differences; this is where food security begins to take on a complexity all its own. In the role of devil's advocate for instance, both FAO and US PL480 consider 'all people at all times' and 'safe' foods; little ambiguity there but what of 'sufficient, nutritious' (FAO and USAID) and '...nutritionally adequate' (USDA) foods? Is there a difference between 'sufficient' and 'adequate' needs? Indeed what are, and who decides on these nutritional needs? Once calculated too, is it a concept in which one-size-fits-all or are there different needs for different people, groups and even regions? The questions do not stop there either. While the USDA focus on 'acceptable foods', the FAO talk of 'food preferences' – in which case: are acceptable and preferential foods the same thing? If so, are these in reference to person's nutritional preferences or in respect of nutritional requirements? Furthermore, while the FAO and USDA talk of 'active and healthy' lives, the USAID and US PL 480 consider 'healthy and productive' lives; the questions then are raised: what are active, healthy or productive lives? How are they determined, provided for and indeed measured? And by whom? Lastly, one does not need to be an expert to realize that the blanket figure of below 2,100 kcal per day, as proposed by the USDA ERS to represent the food insecure, is somewhat problematical. Of course for some these differences might simply be a matter of semantics, For others however, especially theoreticians, statisticians, policy makers and the like, such nuances are unacceptable and need to be properly clarified to the nth degree; especially too if policy is to be predicated on such understandings.

Even with the above expansion of the food security concept, although slightly more involved it still only scratches the surface. For example, the above even though expanded still fails to take into consideration the fact that such analysis defines notions of food security solely at the level of the individual. Factor in issues beyond this simple view and one's eyes are opened to ideas of household and national or regional food insecurity. Taking in yet wider angles we can also factor in the different dimensions of temporary (temporal), chronic (continuous) or cyclical (seasonal) food insecurity too. Indeed, one need only look at the conceptual ideas of the FAO's Vulnerability Information and Mapping Systems (FIVMS) to gain further insight into just how the apparently simple notion of food security becomes exponentially more complex. In efforts to provide fuller comprehension of the subject the FIVIMS model of food security (Figure) provides fifteen 'information domains'. These variables range from the individual to the household to the national, with each in turn further buttressed by supporting factors whose co-integration, co-interactions or codependences are the subject of much research (FIVMS 2008).

### Figure 9.1: Levels of Food Security

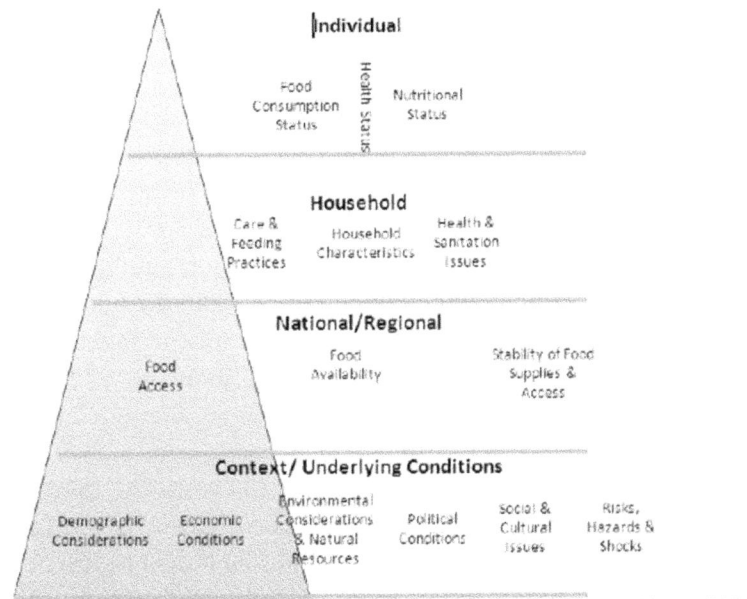

## Back to basics

However, this conceptual posturing is all very well and indeed necessary yet, placing this aside and getting back to basics for the moment, there are very real immediate challenges when it comes to providing food security for all. This is the need to feed a burgeoning population with an annual, finite capacity, food-production system. The two key variables here are population numbers and the food system each of which is affected by many important internal and external influences.

Firstly, taking population numbers and without getting too far drawn into the carrying capacity of the earth (albeit relevant to the discussion), concern over the deleterious effects of unchecked population growth goes back millennia (Freen 1996). In 200 A.D. for example, Roman citizen Quintus Septimus Florence Tertillianus, as quoted by Holland in 1993, had already voiced concerns shared by many of his contemporaries over runaway population growth. In fact the insightful Tertillianus touched on many of the complaints concerning overpopulation which are just as relevant to today's issues of deforestation, biodiversity, farming techniques, increased urbanization and sustainability issues among others (Holland 1993, Johnson 2000, pg1). In Tertillianus own words he described population numbers as being unsustainable and burdensome to the world. Remarkable perhaps (comparatively speaking) when considering global population figures at that time were been estimated at around 200 million (Zhang 2008). Imagine then, after slow and steady growth for the next millennia or so, exponential growth kicked in reflecting compound annual increases of about 0.4 per cent eventually allowing the earth to attain its first one billion inhabitants by about 1800 A.D. From here increases came in ever shorter cycles reaching a second billion in the next 130 years, the third within 30 more years, the fourth within another 15 years, then 12, and 13 years respectively for each subsequent one billion increase (see figure 9.3), (Zhang 2008).

However from about the 17th century onwards mounting concern was already growing over perceived threats from population pressures on the sustainable resources of the day (Malthus 1798, Chalmers 1852). During this period two theories emerged; the first was raised by Thomas Malthus's theory that food resources would ultimately limit population growth while a second concept grew out of increasing collective acknowledgment that the Earth could only support a finite number of people – a naturally carrying capacity. In Malthus' view, the political economist and demographer posited the fundamental inability of the Earth to sustain unchecked population growth. He compellingly argued against the 'perfectibility of man' and formally outlined, perhaps for the first time one of the lasting and well debated theories on population pressures vis-a-vis the Earth's ability to provide (Grigg 1982, Abramitzky and Braggion 2009). More explicitly Malthus suggested that if left unchecked geometric growth of population numbers would eventually overtake food growth with predictable disastrous results (Malthus 1798, Malthus 1803). Malthus' theory led to a phenomenon known as oscillating population numbers or the Malthusian Cycle. Simply put as incomes rose so this allowed for earlier marriages and more children and conversely and just as quickly when the food supply became restricted so marriages and children were deferred leading, in Malthus's view a natural balance of order. Despite the popularity of Malthus's view numerous opponents including Boserup (1965), Johnson (2000) and even Woodruff (1909) among many others suggested that population growth far from being restrictive in terms of agricultural production was in fact a necessity for innovation; and one which ensured supply continued to match demand.

Running parallel to the Malthusian debate is an equally controversial theory of the Earth's natural 'carrying capacity'. Fundamentally similar to Malthus' population versus natural resources theory carrying capacity views the problem from the other side of the argument, which of a limited natural resource supply. The argument presupposes not a maximum number of people the Earth can support but rather an optimum population capacity inferring that the world's population is regulated by the sustainable 'carrying capacity' of the land on which it relies for sustenance (Roughgarden 1979, Daily, Ehrlich et al. 1994). The difficulty here and one that is the subject of much conjecture is the perceived 'limiting' factor to such numbers; whether it be water shortages, food supply or land limitations among numerous others (Lidicker Jr 1962, Grigg 1982, Van Den Bergh and Rietveld 2004, Gilland 2006). Whichever theory one signs up to the core precept remains the subject of much conjecture, as Weisdorf in 2005 offered:

> ... [This] 'chicken-and-egg issue' remains unresolved; did human societies domesticate plants and animals as an adaptive response to population pressure or did domestication give rise to a larger population?" (Weisdorf 2005, p. 566).

Now consider future population trajectories which are on course to yield over nine billion inhabitants by 2050 and concerns of population pressures on both the social fabric and sustainable resources are clearly multiplied considerably (UNDP 2006, USCB 2008).

**Figure 9.2: Historical and Projected Population Trends**

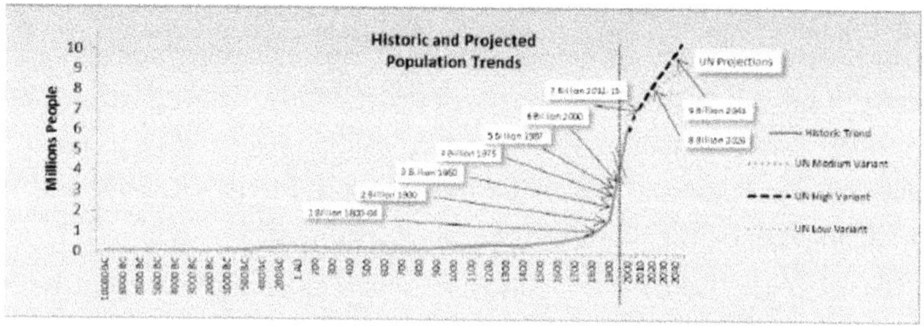

*Source: Based on average past trends as offered by (UN 1973, McEvedy and Jones 1974, Tomlinson 1978, Biraben 1980, Johnson 2000, Haub 2002, USCB 2008, Zhang 2008) and future figures based on UN medium, high and low projections (UNPP 2009).*

So with such burgeoning numbers one could reasonably ask just how have we have fed these people thus far and what are the prospects for the future? This is where the second key variable, the capacity of the food supply, comes into play. It is clear that for present and future population numbers to be sustainably maintained requires a continuous and stable supply of food. Initially, as population numbers began to markedly increase around the early part of the seventeenth century, so growth in the global food supply was largely achieved through reclamation or acquisition of land effectively increasing overall cultivatable acreage. In subsequent periods, from about 1900 A.D. onwards however (in America and the UK at least), available land was becoming harder to acquire so increased productivity was further accomplished through improving crop yields. This was achieved through innovations in both mechanic and bio-technologies as well as improved farming practices; indeed it is no coincidence that population explosion occurred around the time of both the industrial and agricultural revolutions (Johnson 2000, Gardner 2002). These events ultimately paved the way for the mechanization and intensification of agricultural production leading to increased productivity and marked changes in societal dynamics; particularly in such areas as urbanization and off-farm employment growth (Watson 1974, Watson 1983, Johnson 1997, Johnson 2000, Weisdorf 2005).

## Food and population

So, regardless of whether population explosion was the cause or the product of increased agricultural productivity, maintaining sustainable population numbers is predicated on the continuation of an adequate food supply. This notion of 'food security', while it has had a long history in terms of occupying the public conscience (Tertillianus etc.), it has only been around, at the political and institutional level, for a relatively short period of time. At the political level various incarnations of the notion slowly took form around the end of the nineteenth century making its way through several institutional bodies before eventually being taken up by the League of Nations and ultimately the United Nations. Specifically, when the UN launched the FAO in 1945 understanding, quantifying, monitoring and addressing hunger and malnourishment issues around the world became been one of the fledgling body's core activities (FAO 1946, FAO 2013).

## Sufficient food

As mentioned previously, the initial focus of FAO's efforts lay in increasing productivity of the food supply. After many decades however, it became obvious, even to the hardliners, that the reality was much more complicated than this; and herein lies the complexity and difficulty in the food security debate. It was realized that when it came to food production the truth was that supply was regularly providing more than was globally required annually. Currently too, this has been the case now for the last few decades. By way of example we can see that sufficient food is currently produced annually to feed every man, woman and child on the planet with a healthy 2800+ kcals per day; each and every day (Figure 9.3), (FAOSTAT 2013). The problem, as it transpires concern issues of access, availability, stability and the utilization of foods as previously covered. Solutions too it seems, are unlikely to be addressed simply by increasing the supply of food. This is because addressing issues of access, availability, stability and utilization to any degree of satisfaction first requires tackling many fundamental underlying issues of poverty, climate change, equitability, sustainability, subjugation and the numerous other social and political variables – and this is likely going to be a long and drawn out process (Gibson 2012). The sad reality then is that much of this status quo is unlikely to change anytime into the near future; as such much emphasis remains focused on variables of production and supply.

**Figure 9.3 Food Availability (per person per day)**

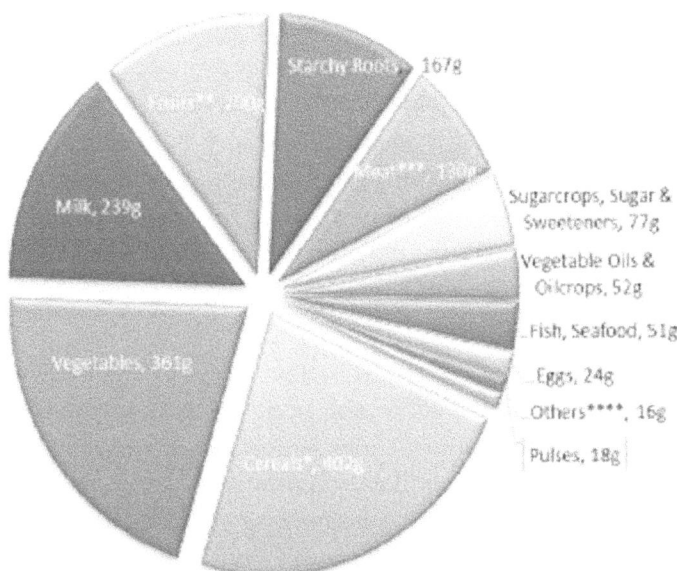

Per Person Daily Food Availability (2009)

*Source: Calculated from latest FAOSTAT Food Balance Sheets of 2009 (FAOSTAT 2013).*

*\*Excluding Beer, \*\*Excluding Wine, \*\*\*Meat including offal and fat, \*\*\*\*Others – Treenuts, Stimulants, Spices.*

## Prospects for the future

So in terms of supply, as alluded to earlier the options are limited and the global community is waking up to the idea that continuing on hitherto unsustainable production practices is no longer viable for both todays' and future population food needs. As a result much research is being carried out in two specific areas of food supply management, these are: production increases and the more efficient and sympathetic use of existing supplies. In the first – production increases, this means considering options like bringing more land under cultivation but, as we have already mentioned is becoming harder to achieve. Another approach is to increase yields through such things as the application of biotechnologies like genetically modified crops or of targeted pest and infestation regimes. Another approach to increasing production is the implementation of improved land management philosophies. This includes better understanding of environmental issues, soil health, improved use of inputs like pesticides and fertilizers etc. as well as an awareness of pollutants, crop rotations, biodiversity and crop varieties among others. In this approach agricultural paradigms are at a crossroads, seeking on the one hand to balance needs previously met through intensive and extensive farming whilst on the other looking at the notion of holistic and more sustainable farming.

This new paradigm seeks to balance many of the issues of current importance whilst also addressing needs of supply. By taking a more sympathetic view to natures' finite larder farmers are actively seeking to live more in harmony with the natural balance of the environment. Sustainability, bio-dynamics, organic farming and food-sovereignty among others, have not only become buzzwords but active movements in which the whole is being seen as greater than the sum of its parts. In this way not only are the food needs of populations being met but many such movements simultaneously aim to address some of the underlying societal indiscretions of inequity, poverty, corporate control, subjugation, human rights and more (IAASTD 2009, Chappell and LaValle 2011). One such approach is organic farming; conceived in 1924 by Rudolf Steiner. It started as a backlash against what was seen at the time as a reduction in soil quality, productivity and ultimately the final taste of the grown products as well as the fear that ultimately:

> "In the long run, the results of attempting to substitute chemical farming for organic farming will very probably prove far more deleterious than has yet become clear (Northbourne 1940, p. 61).

This bio-dynamic movement as it was known at the time was essentially the forerunner to the modern organic movement. The movement was further promoted by the likes of such characters as Lord Northbourne, Walter Ernest Christopher James, Sir Albert Howard and Sir Robert McCarrison, Lady Eve Balfour and of course through the long run Haughley Experiment (over 40 years) (Balfour 1975, Blakemore 2000). From these humble beginnings the world's first organic associations were born and the environmentalist movement thus gained momentum culminating in the seminal work by Rachel Carson in her book 'Silent Spring' in 1962 (Carson 1962).

In another approach the growth of genetic engineering promises to exponentially propagate new breeds of plants and animals with beneficial traits that would see the green and evergreen revolutions pale by comparison. However, from the start there have been problems in the public's perception of modern genetic engineering with much mistrust and general confusion that threatens potential progress. At its most basic, modern genetic engineering (modification) and traditional hybridisation (cross-breeding) are often confused. While both methods aim to produce new crops and livestock that exhibit beneficial traits such as increased yields; increased tolerance to environmental factors; and protection against disease, pests, insecticides and herbicides; both methods however go about it quite differently. Traditional cross-plant and animal breeding happens naturally in nature and involves the cross-breeding of related plant or animal varieties at the non-genetic level (GMES 2007). The big advantage in the plant hybrids is that they are often very fertile producing progeny able to continue the lineage. Animals on the other hand are not able to cross-breed quite so easily and are quite often infertile. Genetic engineering (or genetically modified organisms, GM's or GMO's) of plants and animals etc. on the other hand alters these organisms at the genetic level. This method is often a much quicker process than that which happens naturally and of course desirable traits can potentially be more precisely targeted. Some view the technology as providing limitless potential in therapeutic drug applications, while others can see the benefits of growing more nutritious, long-lasting and naturally pest-resistant crops. However, for many the term 'genetically modified' also comes with menacing connotations conjuring up images of developing new types of hybrids or chimera with potential dire consequences.

However, whatever ones viewpoint, GMO's or transgenic crops are currently grown extensively throughout the world with the potential positive (cautionary notes aside) impact on overall food security. In 2010 for instance GMO's were grown in 29 countries on six continents on over 148 million hectares representing 10 per cent the total 1.5 billion hectares of land used for agriculture. This is a massive 87-fold increase in the 15 years between 1996 to 2010 (James 2010). By way of example of the extent of GM crops, in 2009 by far the four most widely grown crops were:

- Soybeans – of which 77 per cent of the 90 million hectares grown globally were genetically modified;
- Cotton – of which 49 per cent of the 33 million hectares grown globally were genetically modified;
- Maize – of which 26 per cent of the 158 million hectares grown globally were genetically modified;
- Canola – of which 21 per cent of the 31 million hectares grown globally were genetically modified.

Moreover, of the top three producers in 2010 the USA, Brazil and Argentina accounted for over 100 of the 148 million hectares planted globally (James 2010). That said the genetic engineering debate in general and GMO foods in particular is a long and complex one well beyond the scope of this chapter however, despite much mistrust it is worth considering GM crops as one tool in a larger arsenal of food security.

After production increases, in the second approach – the more efficient use of existing supplies, there is much concentration of late in the area of efficiency drives. Some are questioning the competition issues of food for human use and food as substrate for industry; one notable example being the food-biofuel debate. In others much food, fit for human consumption is being diverted for animal feed or is simply being discarded through punishing retail quality control standards. Fish discards too, that proportion of the fisherman's catch which is either over-quota or the wrong species and which is then thrown back often dead, is also an important area of research. There are numerous other examples too; however perhaps the greatest concentration of effort at present reflects the astonishing fact that as much as between 25-50 per cent of all food that is grown for human consumption is in fact being discarded regularly (Stuart 2009, Monier, Mudgal et al. 2010, Parfitt J, Barthel M et al. 2010, Gustavsson, Cederberg et al. 2012).

## Conclusion

From the above discussion it can be seen that the apparent simple notion of food security is quickly complicated by uncomfortable realities. This reality suggests that food security is a broad and compound concept, the outcome of which is determined by the interaction of many political, social and economic considerations. On one side, the march of progress has long been associated with pushing the boundaries of knowledge. This led to the industrial and scientific revolutions which allowed us to do more, achieve more and create more, bringing with it a whole new approach to farming that advocated intensification and industrialization on an unprecedented scale. However, with little consideration of the side effects of this so-called progress, backlash grew in the form of the environmentalist movement. Pioneered by leading lights such as Steiner, Northbourne and Lady Balfour modern alternative agricultural, environmental and sustainable paradigms look to embrace a more practical and sustainable, and in some cases an almost spiritual or ethereal, approach to farming. Treating the earth as a single organism this approach fully embraces the holistic cycle of life (Balfour 1975, Blakemore 2000, Nathoo and Shoveller 2003, Ericksen 2008). Paralleling this has been the social awakening and the increasing trend toward social accountability of a more proactive public in matters of social importance. Collectively these advances mean that today, with increasing ideological convergence, issues of food security are being seen less and less in isolation of much wider public and political aspirations; aspirations which, in the publics' mind at least are inclined to be addressed as a whole and in their entirety (Ruxin 1996, IFAD 2009). Indeed so broad has the catchall of food security become that the FAO, in a rare moment of introspection came to suggest that the phenomenon is now being viewed as belonging anywhere on a continuum of extremes in which on the one hand it represents:

> "... little more than a proxy for chronic poverty...", and on the other, food security:
> 
> "... an all-encompassing definition, which ensures that the concept is morally unimpeachable and politically acceptable, but unrealistically broad" (FAO 2003).

Not surprising then that issues of food security are both advanced and hindered by these expanding aspirations. As they co-exist and sit awkwardly alongside each other the goal is to create harmony between these apparent dichotomies. The challenge, as touched upon earlier is that food security means so many things to so many people. Any solutions therefore will involve a full understanding of just what is in fact included, implied, understood or excluded within both the concept in its narrowest form as well as within the wider food security catchall in general. Only through understanding the complexity and multiplicity of the phenomenon, can stakeholders within the multilateral global development community seek to forge cohesive hegemonic policies to better address such issues. Indeed, until such time as consensus can be found that adequately delineates the phenomenon within these boundaries, current efforts to address the multitude of often divergent threads will no doubt only continue to dilute efforts and confound attempts that once-and-for-all aim to bring these unacceptable figures under control. After all, in reality resources of finance, manpower, political will and of course knowledge are limited that realistically only targeted intervention can be made.

# 10

# Food sovereignty in a post-organic era

*Alana Mann*

## Introduction

In 2009, Bill Gates delivered the keynote address on agriculture at the World Food Prize in Des Moines, Iowa. Founded by Dr Norman Borlaug, father of the Green Revolution, this award "recognises those who have advanced human development by increasing the quality, quantity or availability of food in the world" (World Food Program 2010). At the same time, the first annual Food Sovereignty Prize, created as an alternative to the World Food Prize, was being presented to La Via Campesina at the annual Conference of the Community Food Security Coalition (CFSC). This prize is awarded to "communities, organisations and social movements [that] bring about a more just, healthy and sustainable food system" (Kerssen 2012). These two sets of events, organisations and awards run in parallel and represent competing visions of our global food system, "one engineered in laboratories and centred on global markets, and the other cultivated in the fields, and focused on communities" (Kerssen 2009, p. 1).

This chapter explores the tension between these trends and competing discourses about food, with the objective of interrogating the relationship between the organic food movement and that of food sovereignty. It does so from the perspective of La Via Campesina (the peasant way), the world's most influential transnational agrarian movement. Food sovereignty is defined as "the right of peoples to healthy and culturally appropriate food produced through ecologically sound and sustainable methods, and their right to define their own food and agriculture systems" (La Via Campesina, 2007). The concept of food sovereignty challenges the dominance of agribusinesses and an unjust trade system and promotes an alternative system of small-scale, localised agriculture as a fairer solution to hunger, poverty and climate change.

There is significant overlap between the 'ethical values' and principles of food sovereignty and the organic food movement, including localism, health, ecology, fairness and 'care' in terms of protecting the "health and wellbeing of current and future generations and the environment" (IFOAM, cited in Padel 2009, p. 247).

However important differences underpin the models of production and distribution each movement supports. Central to the concept of food sovereignty is agroecology, a basic principle of which is the diversification of plant species and genetic resources achieved through the integration of crops and livestock to promote biological interactions and synergies within the system (Altieri 2002). This makes it distinct from other alternative farming systems, including organic farming, which may be managed as a monoculture dependent on external biological inputs, particularly when their products become part of the mainstream, corporatised food system. The rise of 'corporate organics' (Johnston et al. 2009) perpetuates for producers the inequality and dependence characteristic of the industrial food system that supporters of the food sovereignty concept aim to subvert.

This chapter describes factors that have contributed to the co-option of organic products by major food companies, such as the exploitation of our 'food fears' and our desires to support local farmers, to create niche markets in the Global North. It begins by introducing an analytical framework for considering the production and consumption of food that defines dominant and alternative discourses in relation to food and agriculture – the productivist, progressive and radical trends. This framework focuses on structure and agency within the agrifood system. Taking an actor-oriented approach, the analysis incorporates a reading of agency that acknowledges the diversity and complexity of the positions from which actors including producers, consumers and retailers approach and make sense of events (Friedland 2008). It critically examines the agency of the politically engaged consumer in making decisions about personal food consumption and discusses the rise of alternative food networks (AFNs) as ways of countering corporate control of the food system. This is followed by an explanation of how radical social movements such as La Via Campesina locate agroecology within a 'politically transformative' peasant movement with the goal of food sovereignty (Holt-Gimènez and Altieri 2013, p. 90). Finally this chapter explores the political opportunities that the food crisis provides for 'tactical, issue-based alliances' (Holt-Gimènez and Shattuck 2011: 135) between the progressive and radical food movements.

## Competing paradigms

There are three major trends in the current food regime, differentiated according to which stakeholders hold power (Holt-Gimènez and Shattuck 2011). The dominant, *corporatist* trend is market-based agriculture driven by transnational agrifood monopolies leveraging policy-makers including governments, the World Trade Organisation (WTO), the World Bank and the International Monetary Fund (IMF). The *progressive* trend is rooted in AFNs that work towards environmentally sustainable and socially just alternatives to the corporatist trend by shortening the food supply chain through direct-selling relationships between farmers and consumers. The *radical* trend, exemplified by food sovereignty, demands systemic changes to the food regime on the basis of rights, entitlements and redistribution.

Alternative paradigms that interrogate food policy in relation to health are a useful supplement to these trends. The *productionist*, *life sciences integrated* and *ecologically integrated* paradigms (Lang and Heasman 2004) focus on agricultural inputs and the application of biotechnology. The productionist model shares with the corporate paradigm a commitment to raising output, intensifying farming, mass processing, mass marketing, homogeneity of product, monocultures and a reliance on chemical and pharmaceutical solutions. The life sciences integrated paradigm claims to deliver environmental health benefits through genetics, biological engineering and laboratory-led solutions while also subscribing to capital-intensive and industrial-scale production to satisfy global demand. In contrast to these top-down, expert-led approaches, the ecologically integrated paradigm is focussed on citizens rather than consumers. In this respect it is closely aligned with the radical food sovereignty movement. It supports agroecological farming methods based on knowledge rather than external inputs, reduces waste and risk, and empowers producers, thus claiming to 'improve links between the land and consumption' (Lang and Heasman 2004: 32). Each paradigm offers a competing, although at times overlapping, views on science, business and consumption.

The 'corporate food regime', as Phillip McMichael (2004) calls it, follows the productionist paradigm. Based on the principles of economic liberalism this dominant, hegemonic and market-based model follows a set of 'rules' written up in free trade agreements, the US Farm Bill and the EU's Common Agricultural Policy (CAP). Reformist elements can be identified within the corporate food regime. These elements aim for a modification of the current system through the provision of social safety nets, Fair Trade schemes and the promotion of organic products (Holt-Giménez and Shattuck 2011). Nonetheless, these solutions are largely technology-driven and rooted in capitalist models of agriculture managed by the same institutions that follow productionist principles. The life sciences integrated paradigm shares with the corporate food regime a foundation in big business and the support of government regulators. The successful rebranding of agribusiness companies such as Monsanto (formerly a chemical producer) as a 'life sciences companies' reflects a disturbing mission creep that is accelerating with calls for a second Green Revolution to prevent future food crises (Robin and Holoch 2010).

This call was prompted by the food price spike of 2007-8, when supporters of the Neo-Malthusian perspective that food insecurity is the result of overpopulation and the attendant drain on global resources spoke of "a new era of dearth, misery and its old companion, vice... set to make a mighty Malthusian comeback" (Hendrix 2011, p. 3). Food sovereignty advocates immediately dismissed claims that aggregate need is outstripping aggregate supply, pointing instead to the dependency and exploitation of the Global South in an unjust trade system as one of the major drivers of hunger and poverty. Claiming that national self-sufficiency through the promotion of domestic markets would release countries from the volatility of international prices by reducing dependence on food imports, former Brazilian president Luiz Inacio Lula da Silva noted "there can be no sovereignty without food sovereignty" (cited in Hendrix 2011, p. 6). Thus food sovereignty simultaneously invokes the power of the state to provide protection and challenges its subordination to the market. It demands that the state provide social support and implement land reform but also pushes the concept of food sovereignty beyond borders into transnational political arenas to pressure governments and educate publics regarding widespread injustices perpetuated by the architects of free trade agreements (FTAs).

La Via Campesina and its allies in the campaign for food sovereignty claim that scientific evidence in favour of the ecologically integrated paradigm, geared towards environmentally sustainable and socially just alternatives to industrial agriculture as a solution to world hunger and supported by the progressive and radical trends, is growing. They draw on evidence that the Green Revolution methods of the 1970s depleted soil organic matter, used excess water and damaged biodiversity, leading many African, Latin America and Asian farmers to turn to agroecological alternatives incorporating traditional and indigenous knowledge in the 1980s (Holt-Gimènez and Altieri 2013). Counter to the arguments of supporters of intensive, industrial farming these methods are capable of producing crops with high productivity and resilience but with lower environmental impacts, as revealed in the path-breaking analysis of food systems carried out by the UN's International Assessment of Agricultural Knowledge, Science and Technology for Development (IAASTD 2009) following the 2008 food crisis. The Assessment shares the view of the radical movements that small-scale agriculture is a moral enterprise at odds with a market-driven globalisation manifested in corporate concentration, highlighting the deprivations of small-scale farmers and particularly women. It claims that "policy and institutional failure has limited the use of sustainable practices; it could also be argued that this is the underlying reason why people are malnourished, farmers are poor and the price of food is rising" (Mulvany 2008, p. 26). This supports La Via Campesina's demands for systemic changes to the food regime on the basis of rights, entitlements and redistribution – principles central to the concept of food sovereignty.

For La Via Campesina questions of food safety, quality and human rights have given rise to a moral economy of food (Slater 2004) that is offered as an alternative to a model of globalisation based on capitalism and development. Food sovereignty aims to return control of food production and consumption to citizens through democratic processes rooted in localised food systems. The food sovereignty movement aims to "build close links with people living in the urban centres in order to provide them with healthy food from people to people, without the destructive interference of transnational corporations" (La Via Campesina, 2009a, p. 57). In this respect, the discourse of food sovereignty directly challenges an existing global food regime characterised by 'distancing and durability' (Friedmann cited in Weis 2007, p.14) and resonates with consumers in the Global North where "food has been taken for granted" (Marsden 1997, p.169).

For supporters of the productionist or corporate food regime, solutions to the ongoing global food crisis include the expansion of GMO cultivation, liberalised markets and internationally sourced food aid financed largely by the World Bank's Global Agriculture and Food Security Programme, a multilateral trust fund set up by the US, Canada,

Spain and the Bill and Melinda Gates Foundation (Holt-Gimènez and Shattuck 2011). Coupled with a series of industry-NGOs and public-private partnerships (PPPs), the Programme represents a continuation of the 'development project'. The transformation demanded by supporters of the radical trend is directly opposed to this paradigm. The aim of the radical social movements is to convince stakeholders that 'food embodies social, cultural and ecological values over and above its material value' (McMichael 2008, p. 49). This discourse offers a revitalised politics of agrarian citizenship where the rural is the civil base in a new political economy of representation (Wittman 2009). Conventional terms – sovereignty, agrarian reform, citizenship and rights – gain new meaning in this economy. Realising and sustaining these new subjectivities requires solidarity and the formation of a collective identity. This collective is based on the recognition of different struggles being carried out in response to policies that impact negatively on farmers world-wide in the form of low crop prices, high subsidies and the disappearance of family farms. Thus, the tensions surrounding food production lie not in conflicts between governments but within the models of production and rural development operating in both the Global North and South. For La Via Campesina, the question is fundamentally social – who should provide food and through what relationships? Whose livelihoods should be protected?

According to the radical movements, 'the enemy is the [neoliberal] model' and a transition away from this model is needed (Rosset 2006, p. 309). La Via Campesina calls on "all those responsible in governments to step out of the 'neoliberal model' and to have the courage to seek an alternative path of cooperation with social justice and mutual assistance" (La Via Campesina 2003). It argues that market-led agrarian reform (MLAR) is inadequate in highly unequal societies where reform processes favour large landowners whose land often lies unused and fulfils no social and economic benefit. It challenges the assumptions on which the rural initiatives of development agencies such as the World

Bank are based, for example that agriculture is the main source of economic growth, and therefore increasing productivity is the solution to poverty. These initiatives require the liberalisation of markets, the inclusion of agriculture in multilateral FTAs, the strengthening of private companies, the privatisation of sectors previously controlled by the state, investments in biotechnology and support for diversification of export agriculture (Patel 2006). La Via Campesina challenges these 'common sense' solutions, calling for the G8 to 'clean up their own mess instead of dictating to poor countries what to do' (La Via Campesina, 2009b). It emphasises that the solution to the global food crisis does not exist in MLAR, which has led to the dispossession and migration of a significant proportion of rural peasants.

Market failure is the basis of the radical movements' response to economic arguments regarding how to address global poverty and hunger. According to the discourse of food sovereignty, the neoliberal model of production is based on the principle of overproduction by the 'grain-livestock complex' in the temperate world (Weis 2007, p. 86-7). The US, specifically, has achieved 'tremendous productivity gains, exported surpluses, industrial innovations and the rise of its agro-TNCs' (ibid) resulting in extreme concentration of production and insurmountable inequality among producers. As a result, farmers throughout the Global North are trapped in a 'cost-price squeeze' while distorted competition from cheap exports has ruined largely unsubsidised farmer livelihoods in the Global South. In contesting the food enterprise discourse of the neoliberal model, the radical movements extend the rights-frame of food sovereignty to the promotion of the peasant farmer as a viable economic entity (Holt-Gimènez and Shattuck, 2011). This active engagement in the construction and deployment of economic frames is essential in countering what Marc Edelman (2008) considers to be hegemonic conceptions about the rights of economic actors in the market that undermine the capacity of human rights and environmental frames. The G8's economic mandate, for example, is less vulnerable to civil society tactics such as naming and shaming than national governments. La Via Campesina has presented a clear prognostic frame regarding the roots of the global food crisis in financial markets:

> This current food crisis is the result of many years of deregulation of agricultural markets, the privatisation of state regulatory bodies and the dumping of agricultural products on the markets of developing countries. According to the FAO, liberalised markets have attracted huge cash flows that seek to speculate on agricultural products on the 'futures' markets and other financial instruments (Saragih 2008).

The rift between the food enterprise and food security discourses of the neoliberal model, on the one hand, and that of food sovereignty, on the other, can be traced to modern theories of economic development. Food sovereignty is premised on 'justice between all economic actors' achieved by agricultural trade based on 'relationships of equality, cooperation and fair exchange' (La Via Campesina 2009a, p. 61). In contrast, according to McMichael (2004, p. 57), the relationship between the industrial or corporate food regime and the current project of global development, represented by the WTO's Doha Development Round, has redefined and institutionalised food security as an 'inter-nationally managed market relation'. The capitalisation of nature is reflected in the discourse of 'sustainable development' that aims to reconcile the economy and ecology in a discourse Harvey (1996, p. 382) refers to as 'ecological modernisation' Despite its 'radical-populist edge', ecological modernisation is reliant on scientific evidence that reinforces rather than challenges the capitalist economic system. Based on the premise that economic activity produces environmental harm, ecological modernisation draws on the discourse of sustainable development as an ideal where the economy grows with respect for natural limits. This 'win-win' discourse has been appropriated by TNCs that use it as a justification to obtain resources. According to this discourse environmental management becomes reliant on government policy and therefore subject to the hierarchy of powers; looking after the environment is good for business and "only minor adjustments to the market system are needed to launch an era of environmentally sound development, hiding the fact that the economic framework itself cannot hope to accommodate environmental considerations without substantial reform" (Escobar 1995, p. 197). Food sovereignty has emerged as an 'alternative principle to the productivism and quantitative measures of food security identified with the monetised transactions of the market system ... premised on a farmer-driven agriculture that is the key to food-secure relations of environmental and social sustainability' (McMichael 2004, p. 58). It counters traditional discourses of development that formulate a North-South binary and recognises that small-farm households represent 40 per cent of humanity (Van der Ploeg, 2008, p. xiv).

**Peasant activism**

Not only are peasantries an integral part of contemporary society but provision for peasant forms of rural and agricultural development may be the only solution to hunger and poverty. Yet peasants are subject to 'invisibility' within the corporate food regime (Van der Ploeg 2008, p. 269). Following Hardt and Negri (2000), Van der Ploeg defines this 'empire' as 'an ordering principle' that governs the food system through ecological and socio-economic exploitation. In the food empire expansion and hierarchical control has led to the creation of new material symbolic orders and a segmentation of agriculture. In his longitudinal studies of Peruvian, Dutch and Italian farming communities, Van der Ploeg identifies a lack of attention to the distinctiveness of peasant agriculture in comparison to other forms of farming. The framing of farmers as either capitalist or peasant, he argues, is too narrow. Moreover, peasant modes of farming are not confined to the Global South. Although conditions and outcomes are different in the Global North, notions of peasant farming are re-emerging as a means of understanding transitions in the European countryside.

According to Van der Ploeg (n.d) the peasant condition and peasant modes of farming can be seen as a 'struggle for autonomy' in a context of marginalisation. The aim of peasant agriculture is to create and develop 'a self-controlled and self-managed resource base' that facilitates 'forms of co-production of man and nature that interact with the market, allow for survival and for further prospects and feed back into and strengthen the resource base, improve the process of co-production, enlarge autonomy and, thus, reduce dependency' (p.23). Van der Ploeg describes a process of 'repeasantisation'. Distinguishing between peasant, entrepreneurial and capitalist forms of agriculture, he defines repeasantisation as "a modern expression of the fight for autonomy and survival in a context of deprivation and dependency" (p.7).

Historically, this fight has been characterised as reflex reactions to economic or political oppression that take the form of rebellions or volatile uprisings. In these constructions, the peasant rebel is portrayed as lacking individual will and agency, existing only as a part of a faceless class. Ethnographer James C. Scott challenged this representation of peasant passivity in *Weapons of the Weak* (1985) in which he argued that 'everyday forms of peasant resistance' on a local level in fact complement highly visible examples of critical collaboration and confrontation. His study of a Malaysian village challenges the depiction of peasants as 'anonymous contributors to statistics on conscription, taxes, labour, migration, land holdings and crop production' (1985, p.29), Rather, his study foregrounded the existence of a subversive resistance. Tactics such as 'foot dragging, dissimulation, false compliance, pilfering, feigned ignorance, slander, arson [and] sabotage' denied the claims of authorities while advancing those of others (p.29).These tactics – informal, covert and rarely co-ordinated – are the ordinary weapons of powerless groups.

> Everyday forms of resistance make no headlines. But just as millions of anthozoan polyps create, willy-nilly, a coral reef, so do the multiple acts of peasant insubordination and evasion create political and economic barrier reefs of their own (Scott 1985, p. xvii).

This interpretation of resistance challenges the Gramscian concept of hegemony, arguing as it does that subordinate classes are equipped to 'penetrate and demystify the prevailing ideology' through an understanding that goes beyond the limited, reformist language of the 'trade union consciousness'. The resistant peasant is following a 'hidden transcript' that recognises 'the necessity of routine and pragmatic submission to the compulsion of economic relations, as well as the realities of coercion' (Scott 1985, p. 17-18). Everyday forms of peasant resistance begin close to the ground, rooted as they are in 'the meaningful realities of daily experience' (p.348). At this level, enemies are real people rather than faceless multinationals, global governing institutions such as the WTO, or historical factors. The values promoted are at hand and familiar, the goals modest and concrete rather than ideological abstractions.

As Scott's analysis suggests, earlier literature portrayed peasants as passive victims without agency. Questions of structure and agency have long occupied sociologists of agriculture and food. Towards the end of the last century, the focus was on the crisis of the nation-state, the 'globalisation project' and the consequence of these trends on food production and consumption. The liberalisation of markets, growth of TNCs, advances in biotechnical solutions, exploitation of labour as well as the encroachment of regulations in both the public and private spheres have provoked resistance on a range of battlefields where 'resistance is encountered in a wide range of heterogeneous and increasingly interlinked practices through which the peasantry constitutes itself as *distinctly different*' (Van Der Ploeg 2008, p. 265). This difference is a source of identity solidarity for the transnational social movement of peasant organisation that is La Via Campesina.

## La Via Campesina – The evolution of a transnational social movement

The immediate roots of La Via Campesina can be traced back to 1992 when eight peasant and family farmer organisations from Central

America, the Caribbean, North America and Europe met in Managua, Nicaragua, during the Second Congress of the National Union of Farmers and Ranchers (Desmarais 2007). At this time the only major international farmers' organisation was the International Federation of Agricultural Producers (IFAP), founded in 1946 with the objective of preventing food shortages such as those experienced during the Great Depression of the 1930s. At Managua, farmers shared their experiences of the impact of state and international policies on agriculture and rural communities and agreed that the current economic model based on free trade, low prices and industrial agriculture was unjust and unsustainable. An alternative model was needed, they argued, one that gave peasant farmers a central role in developing the rural and food policies that impact them directly.

The following year more than 70 leaders of farmers' movements met in Mons, Belgium in May for the First International Conference of La Vía Campesina. It was here that divergences of opinion between IFAP and small producers on trade liberalisation were formally articulated. IFAP was accused of representing the interests of larger farmers in industrialised countries over those of small producers. It was also seen, in some countries, as supporting agricultural policies that were in fact detrimental to peasant agriculture. Moreover, the promotion of policies aimed at enabling farmers in the Global South to catch-up with the Global North was widely viewed as paternalistic and ignorant of local priorities and needs.

A further catalyst for worldwide solidarity between farmers' organisations arrived with the creation of the WTO through the Marrakesh Agreement in 1994, after the conclusion of the Uruguay Round of General Agreement of Tariffs and Trade (GATT). The establishment of this global governing body charged with liberalising trade marked an irrevocable change in the relationship between rural producers and national governments. From now on domestic agricultural policies would become secondary to international agreements. Explicit opposition to neo-liberal trade policies, and the WTO in particular, became a common ground for a transnational movement of small-scale farmers fighting inequities in international trade agreements, labour exploitation, the application of biotechnology, environmental pollution and the implementation of "bureaucratised systems for the measurement and regulation of product quality and food safety" (cited in Van der Ploeg 2008, p. 267).

Today, La Via Campesina claims to be the world's largest social movement with 170 member organisations in 70 countries representing approximately 200,000 small-holder farmers, fisher folk, seasonal workers and landless peasants. Their identity solidarity is based on the premise that all small farmers are subject to the same impacts of neoliberalism's structural violence, "the unprecedented concentrations of wealth and power and the rapid destruction of life-ways and livelihoods, eco-systems and species" (Reitan 2007, p. 16). La Via Campesina claims that:

> Neoliberal policies prioritise international trade, and not food for the people. They haven't contributed at all to hunger eradication in the world. On the contrary, they have increased the peoples' dependence on agricultural imports, and have strengthened the industrialisation of agriculture, thus jeopardising the genetic, cultural and environmental heritage of our planet, as well as our health. They have forced hundreds of millions of farmers to give up their traditional agricultural practices, to rural exodus or to emigration (La Via Campesina 2006).

In 1996 La Via Campesina introduced the term 'food sovereignty' at the World Food Summit as a foil to the notion of food security. Food sovereignty exposes questions around the 'how' of the food system – its social control. Patel states that "as far as the terms of food security go, it is entirely possible for people to be food secure in prison or under a dictatorship ... the absence of specification about how food security should come about was diplomatic good sense" (Patel 2009, p. 665). La

Via Campesina addresses this omission by constructing the following relationship between the two concepts:

> Long-term food security depends on those who produce food and care for the natural environment. As the stewards of food producing resources [peasant farmers] hold the following principles as the necessary foundation for achieving food security: food is a basic human right. This right can only be realised in a system where food sovereignty is guaranteed. Food sovereignty is the right of each nation to maintain and develop its own capacity to produce its basic foods respecting cultural and productive diversity. We have the right to produce our own food in our territory. Food sovereignty is a precondition to genuine food security (La Via Campesina 1996).

Through its adroit use of the discourse of securitisation to which states have made commitments, this interpretation introduces questions regarding the power relations inherent in the food system.

The Declaration of Nyèlèni (2007), devised at the Forum for Food Sovereignty in Mali in 2007, provides the most recent and comprehensive definition of food sovereignty. It emphasises social relations by "putting the aspirations and needs of those who produce, distribute and consume food at the heart of food systems and policies rather than the demands of markets and corporations". It "prioritises local and national economies and markets and empowers peasant and family farmer-driven agriculture ... fishing, pastoralist-led grazing, and food production, distribution and consumption based on environmental, social and economic sustainability" within a human rights framework that "ensures that the rights to use and manage lands, territories, waters, seeds, livestock and biodiversity are in the hands of those of us who produce food'. In doing so it "implies new social relations free of oppression and inequality between men and women, people, racial groups, social and economic classes and generations" (La Via Campesina 2007).

Supporters of food sovereignty seek to relocate power back to producers and consumers. In the corporate food system "those that control loans, materials supply, the dissemination of new technologies, such as transgenic products, on the one hand, and those that control national and international product warehousing systems, transportation, distribution and retail sales to the consumer, on the other, have real power", according to La Via Campesina (cited in McMichael 2011, p. 1). The transformative change that the radical social movements demand requires the dismantling of a global food regime controlled by powerful agrifood monopolies.

## From seed to shelf, from farm to fork

In the contemporary global food system major buyers and sellers with the capital to dominate transportation and distribution currently connect millions of producers and consumers in a supply model resembling an hourglass whereby a small number of agribusiness companies act as mediators between millions of producers and billions of consumers. As Raj Patel says, "the small fish have been devoured by the Leviathans of distribution and supply" (2007, p. 12). These 'invisible giants' (Kneen, 2011) own and control all stages of the food production and distribution process. This form of supply chain management reduces competition, resulting in the rise of major food firms engaged in joint ventures and strategic alliances. The giant transnational bulk commodity companies - Archer Daniels Midland (ADM), Bunge, Cargill, and (Louis) Dreyfus, collectively known as the ABCD group based on their initials -manage overlapping food, fuel and feed complexes. They are engaged in cluster relationships with biotechnology and seed companies including Monsanto, Syngenta, DuPont, Aventis and Dow (Murphy et al: 2012; see also Hendrickson et al, 2008; Hendrickson and Heffernan 2002; and Lang 2003). In 2008 while hunger riots took place in 40 countries Cargill made record profits of nearly $US4 billion (Davis 2009). In response to Bunge's announcement of soaring profits in 2010, CEO Alberto Weisser confessed:

> I hate to say that we benefit ... what we have done is a very deliberate strategy to build a global network of systems to be one of the companies who can provide food when it's necessary (Meyer 2010).

The food chain clusters profit from a trade regime that increases capital mobility, reduces costs and creates a global division of labour. Monopolies expose producers to global differentiations in costs and "control the food system from gene to supermarket shelf" (Heffernan, cited in McMichael 200, p. 25). Consumers are encouraged to rely increasingly on the expert knowledge of food manufacturers, labellers and processors. The corporatised food regime makes discursive claims regarding bio-technological solutions, sustainable agriculture, efficiency, small government and competitive parity between the Global North and South (McMichael 2000) while in reality the capital intensive, privatised model of agriculture and the purchasing of inputs from outside the farm contribute little to rural livelihoods (Hendrickson et al, 2008).

A redistribution of power is occurring with the parallel consolidation of major food retailers who integrate backwards in the food system by forging relationships with food chain clusters. Commodity companies such as the ABCDs are increasingly losing their share of world trade in food and fibre to global retailers such as Wal-Mart, Carrefour and Tesco. These retailers are responding to consumer tastes and preferences for processed and value-added products that have eclipsed bulk commodities in global trade. 'Category captains' such as Unilever create strong stables of brands and dominate entire ranges of products (Hendrickson 2008, p. 28). Supermarkets have developed "integrated supply chains, from farm to fork, under the prescriptive orchestration of the retailers" (Kjaernes et al, 2007, p. 133), a global trend which results in counter-seasonal produce and a transformation in the range and variety of food available. This consolidation of food retailers increases their power to set the prices paid to food producers, thus reducing the latter's power to respond to changes.

## Translating risk into marketing opportunities: corporate organics

The "geographies of trust" (Freidberg 2004, p. 11) between food producers and consumers have shifted with advances in transport and communication. These advances have facilitated a distancing within the food chain that conceals potentially unsafe, unethical and environmentally damaging production practices from consumers. The central connection between producers and consumers, impersonally labelled the market-exchange relationship, is based on trust. In Ulrich Beck's (1999) risk society, in which governments have failed to offer adequate protections to citizens, consumption has become a new form of 'subpolitics' that empowers the individual to take more responsibility for their choices. As citizens respond to the increasing complexity of the risk society, they seek to recapture power within the food system and regain a more intimate relationship with what has become an industrial commodity. The consumption of 'free range' and organic products, for example, becomes a means for consumers to regain control over food through self-regulation. Thus, the distinction between public and private is blurred in that the state may no longer be the dominant actor in multi-layered networks of decision-making.

With technological advances in food production and processing, come increased risk and complexity. This has led to new systems of governance, trade and traceability schemes that enable the tracking of a product from farm to plate. These complex systems reflect the struggle between globalisation and localisation. Food governance has been complicated by revolutions not only in food supply and distribution leading to longer supply chains, but in developments in our understanding of nutrition and health, shifts in political ideologies, changing consumption habits, population demographics, and the rise of new stakeholders including civil society actors such as the radical social movements (Lang and Heasman: 2004).

Risk is constructed and translated by different actors along the food production-consumption chain in various ways. Political mobilisations around food issues that represent 'collective expressions of mistrust' (Kjaernes et al, 2007, p. 46) have driven demands for transparency and the participation of a wider range of actors in policy-making. This has led to the emergence of regulatory agencies and independent, scientific advisory bodies to restore public trust in governance. In the formulation of food policy, complex trust relationships exist between these bodies, consumers and producers. There has been devolution of individual responsibility to corporate food retailers through the enforcement of supplier standards, which can have severe repercussions for poorer exporting nations. State deregulation passes governance of the food chain to retailers, who exercise technical, discursive and social power over producers who must abide by standards to compete with other suppliers. The exercise of this power can be interpreted as a new form of 'postcoloniality' (Freidberg 2004) that passes additional costs and risks to producers, particularly exporters in the Global South. Small farmers may still be kept hostage to market forces, as well as regulatory measures such as certification seals. Meanwhile, industry constructs risk favourably to deflect consumer attention away from food safety and towards the capacity of technology to save humanity.

As food safety scares create crises of confidence and undermine trust in the corporate food system they represent political opportunities for progressive and radical food movements. According to a 2004 survey commissioned by Kellogg (Bostrom 2005), the popularity of organic food can be largely attributed to outbreaks of diseases such as Bovine Spongiform Encephalopathy (BSE) and salmonella as well as to more general concerns over the health impacts of genetically modified organisms (GMOs) and pesticide use. However corporations are just as strategic in exploiting crises. In 'capturing then codifying consumer concerns', manufacturers and retailers translate food fear into food product (Blay-Palmer 2008, p. 135). As consumers tend to compartmentalise crises, they do not engage in the systemic issues of the food system. They interpret food scares as episodic events, enabling food distributors, retailers and marketers to isolate and manage problem areas before the bigger picture, which exposes systemic flaws, can be seen. Corporate food companies carry out 'strategic convergence' in response to food safety concerns by launching product quality measures and new, territorially identified products such as Waitrose's 'Welsh Organic Lamb' (Goodman 2004, p. 9). This raises the question of a new, porous border that has developed between the marketing of 'conventional' and 'alternative' food (Goodman and Goodman 2007).

The commandeering of organic labels by conventional agribusiness firms is a 'post-organic' response to the mainstreaming of organic products within the corporate food system. Through labelling schemes developed to provide alternatives to conventional food supply chains that which was an 'anti-institutional' movement has become institutionalised (Goodman and Goodman 2001). As products enter global retail chains, they are marketed as niche foods (Watts 2005) thereby losing their power to transform the industrial food system. 'Corporate organics' (Johnston et al, 2009) describes those elements of the organic food sector that have moved beyond small scale direct-selling models of distribution to a corporate model of factory farming supplying distant supermarkets. The Leviathans of the supply chain have consumed many original organic companies. In the US, organic brands Lightlife, The Organic Cow and Back to Nature, for example, have been acquired by Heinz, ConAgra and Kraft Foods (Johnston et al, 2009). Accordingly these brands have become part of long-distance commodity chains subject to centralised corporate control. Ironically, the institutionalisation of organic agriculture through certification standards and labelling systems, along with highly successful campaigns promoting the health and environmental benefits of organics for consumers and producers, have helped draw corporates to the sector (Goodman and Goodman 2001). For retailers, demonstrating accountability throughout the food supply chain from grower to market fulfils their corporate social responsibility objectives while simultaneously carving out competitive space.

Resistance to this trend has led to calls for 'food democracy' represented by community supported agriculture (CSA) schemes that bypass major retailers. This represents a schism in the progressive food movements. Food democracy supports organic agricultural techniques but also focuses on fair wages and better living conditions for farm workers, themes that resonate with food sovereignty advocates. It resists cooperation with transnational corporations and major supermarkets and offers consumers ways of expressing dissatisfaction with the conventional food system.

## Consuming subjects versus consuming citizens

The modern supermarket is the dominant contemporary symbol of capitalist penetration into the food system with its endless array of processed, packaged and scentless products. The conventional, industrial global food system 'tends to engage with food production as if food were a commodity like cars or widgets' placing emphasis on quantity and large-scale production, centralisation and technology (Blay-Palmer 2008, pp. 2-3). Marketing discourse, a powerful tool of the market economy, addresses citizens as 'consuming subjects' for whom it constructs attitudes to food products (Blay-Palmer 2008). Advertising campaigns aim to capture the connection between producer and consumer.

> Origin, quality, authenticity, freshness and specificity of products, and of associated ways of producing, processing and marketing, are clearly articulated in order to attract consumers and to communicate the distinction embodied in food – the distinction that 'passes' to the consumers themselves (and to the act of consumption) (Van Der Ploeg 2008, p. 279).

Disturbingly, food democracy themes pervade the marketing discourse that promotes corporate organics. Themes including enduring connection to family farms, idealistic portrays of agrarian life and small-scale 'humble beginnings' (Johnston et al, 2009, p. 519). Consumers must critically interrogate these claims in a policy environment that 'devolves regulatory responsibility to consumers via their dietary choices' (Guthman 2007: 264). This is a form of political consumerism, defined as "actions by people who make choices among producers and products with the goal of changing objectionable institutional or market practices ... based on attitudes and values regarding issues of justice, fairness, or noneconomic issues that concern personal and family well-being and ethical or political assessment of favourable and unfavourable business and government practice" (Micheletti 2003, p. 2).

The progressive social movements are genuinely concerned with the relations of consumption vis-à-vis the relations of production (Holt-Gimènez and Shattuck 2011). They seek to rebuild rural-urban relationships, favouring fresh, locally-sourced food that returns value to the producer through AFNs and CSA schemes. These initiatives aim to reconnect with and reconstitute local human, cultural and land ecologies as a means of creating and connecting new spaces and models for engaging the public in debates over environmental sustainability, social justice and economic viability (Gottlieb and Fisher 1996).

In the US the origins of the progressive food movement date back to the 1920s when environmental justice movements raised questions of food production, distribution and access. Concepts such as the 'overcity', 'cosmopolitan city of scale' and 'garden city' suggested the need for a balance between urban and rural life (Gottlieb and Fisher, 1996, p.24). Later policy narrowed to a focus on food safety, attributed to the lack of an integrated approach between environmental and food security advocates. The sustainable agriculture movement, referred to prior to the 1980s as the 'organic farming movement', focussed on the role of producers, drawing attention to the growing influence of agribusiness and the decline of rural communities.

Today, Food Policy Councils in the US and Canada and Associations for the Maintenance of Smallholder Agriculture (AMAP) in Europe promote AFN initiatives in the Global North while NGOs support many groups in the Global South through schemes including Fair Trade (Holt-

Gimènez 2011, p. 126). In Europe, the Slow Food Movement founded by Carlo Petrini in Italy in 1986 promotes biodiversity, sustainable farming, local production and food heritage from the perspective of the gourmand and the conservationist. With 80,000 members globally, the organisation has established the University of Gastronomic Sciences and Terre Madre events supported by famed international chefs, environmentalists and His Royal Highness, the Prince of Wales (Petrini and Padovani 2006). These groups have emerged in response to the global corporate food regime and sympathise with elements of the food sovereignty platform including the right to food, agroecological production and better wages for agricultural workers.

In equating food with social relations and farmer livelihoods, these progressive movements have much in common with the more radical politics of La Via Campesina. They are concerned that the corporate food system generates 'food from nowhere' (Bovè and Dufour 2002, p. 55) swapped between countries, generating countless 'food miles' which intensifies the already alarming impacts of industrial agriculture on the environment (McMichael, 2007). AFNs create "authentic social, economic and ecological relationships between all actors in a food system" (Hendrickson and Heffernan 2002, p. 361) in which consumers are recognised as actors in mutually constituted food circuits. Proximate relationships with growers, built on trust, enable consumers to bypass the political economy of the industrial food system. Shortening food supply chains either physically, by buying directly from producers, or conceptually, by opting for Fair Trade products, achieves this goal for political consumers. Through their changing consumption practices, for example, respecting the boundaries of local 'food-sheds' (Getz 1991) by purchasing at farmers' markets, consumers are expressing their dissatisfaction with corporate control of the food system. Products identifiable with a specific geographic area promote relationships built on trust and allow food to be relocated in the life-world in a way that conventional urban retail experiences cannot, for while "advertising (promoting brands) can create the illusion of connection, it is only within the context of integrated relationships that authenticity can be developed" (Hendrickson and Heffernan 2002, p. 361). Yet these claims must also be interrogated.

## Interrogating the 'Local'

The objectives of those who frame resistance to the global in terms of reconstructing the local should be viewed critically, especially when they make claims regarding environmental sustainability, economic viability and social justice. David and Michael Goodman (2007) challenge the normative assumptions entrenched in localist discourse and maintain that small-scale organic farming operations in fact often follow capitalist market logic. Localist discourse emphasises social, place-based, community values where "the local is framed as a social space, where new economic norms and institutions incorporate ethical norms that are allowed to grow and flourish" (Goodman and Goodman 2007, p. 29) but frequently privileges economic discourse over social justice issues. For example, a study of local food initiatives in the US (Allen and Hinrich, 2007) identifies in campaign materials an emphasis on economic support of farmers and associated businesses where the aim is to keep money in the community rather than raise awareness of the inequalities of the global food system. Therefore, understandings of the local do not necessarily translate into socially just alternatives.

Localism and regulated consumer choice part of the neoliberal discursive field, and as such engagement with AFNs can represent a form of 'high politics' (Beck 2000, p. 122) through which consumers depict themselves as *non*-political. Localism may be a reaction against globalisation and yet still present inequalities. Harvey (1996) argues that "transformations of space, place, and environment are neither neutral nor innocent with respect to practices of domination and control", describing such transformations as "fundamental framing decisions – replete with multiple possibilities – that govern the conditions (often oppressive) over how lives can be lived" (Harvey 1996, p. 44). To develop these authentic relationships within the structure of everyday consumption praxis is challenging and requires consumers to appreciate the domestic, public, civic and ecological issues involved in making food choices (Murdoch et al, 2000). It requires them to pose the question "sustainability for whom?" (Seyfang 2006, p. 386).

## A Battle of Farming Philosophies

A central principle of food sovereignty is the achievement of sustainable livelihoods through co-production between man and nature. Organic farmers share the desire to base agriculture on living ecological systems and cycles, working with them, emulating them and helping sustain them (IFOAM cited in Padel 2009). The two philosophies depart, however, on the issue of monocultural farming, described by Michael Pollan as "the root of virtually every problem that bedevils the modern farmer" (2001, p. 225). The cultivation of a single plant species, characteristic of industrial agriculture, not only reduces biodiversity but threatens the livelihoods of those dependent on a single crop.

Small-holder farmers' engagement in pluriactivity – involvement in activities unrelated to agriculture – is one of the many ways in which land is improved. Pluriactivity is thus a form of risk management in that the impact of the failure of a single crop and the longer-term negative environmental impacts of farming is reduced and biodiversity maintained. Miguel Alteri (2010, p. 259) argues that not only are small-scale, pluriactive farms more productive and resource-conserving than large monocultural set-ups, they represent a 'sanctuary of agrobiodiversity free of GMOs', are 'more resilient to climate change' and create carbon stores. Agroecology, defined "the application of ecological concepts and principles to the design and management of sustainable agricultural ecosystems" (Altieri and Toldedo 2010, p. 254-5), provides the foundation of an "epistemological, technical and social revolution ... from below" (p.587). This revolution has been linked with progressive political developments in Ecuador, Bolivia and Brazil. For opponents of the corporate food system agroecology represents a solution to declining local economies and rural unemployment. As a response to a production model that relies on fertilizers, seeds and herbicides as well as high levels of regulation and certification, agroecology is affordable and sustainable. In its 2012 *Surin Declaration* La Via Campesina declares:

> Agroecology is giving a new meaning to the struggle for agrarian reform to empower the people. The landless farmers who fought to reclaim back their land, and those who received land through land reform programs in Brazil and Zimbabwe, are implementing agroecology as a tool to defend and sustain their farming, not only for their families but to provide healthier food for the community. Therefore, land reform, together with agroecology, has become the contribution of peasant and family farmers to give better and healthier food to our societies (*Surin Declaration*, 2012).

## Convergence between the organic and food sovereignty movements

An assessment of the relationship between the concept of food sovereignty and the organic movement reveals areas of strategic convergence. Organic and food sovereignty movements are in fact the "arms and legs of the same movement" (Holt-Gimènez 2010, p. 4). Supporters of food sovereignty and organics agree that "we are embedded in a global food system structured around a market economy that is geared to the proliferation of commodities and the destruction of the local…faced with transnational agribusinesses whose desire to extend and consolidate their global reach implies the homogenisation of our food, our communities, and our landscapes" (Kloppenburg et al., 1996, p. 36).

A significant barrier to the creation of a unified counter-movement that can offer realistic alternatives to the existing food regime is that the organic movement is embedded in capitalist models of production and operates within the constraints of conventional agriculture. The method remains compatible with monocultures; if this does not change then the most the organic movement might achieve through mass uptake of organics is "denting the profits of pesticide companies" (Patel 2009, p. 306). Without a dramatic change in the organic movement's framing of the 'structures of provision' (Seyfang 2006, p.385) unification with the radical movements cannot be realised.

Secondly, while the virtues of organic food are increasingly understood in the Global North any attempt to link the interests of middle-class consumers – food safety, the environment, animal welfare and taste – and those of peasant farmers engaged in a struggle for equality in social relations and fundamental changes in the distribution of resources, remains challenging. Food sovereignty incorporates claims for the defence of cultural difference and territories. These claims challenge base inequalities and demand protection from food dumping, cultural imperialism, biotechnology and other activities that threaten the natural connections between producers and consumers. They go beyond the primacy of individual property rights to a model of land reform based on the special nature of agriculture and its multifunctionality, and are focused on preserving landscapes, protecting livelihoods and valuing rural traditions (Rosset 2006). Therefore, reform embraces the comprehensive revision of agricultural systems to favour the production and marketing of small farm produce. According to La Via Campesina, the external constraints imposed by international trade agreements not only lessen economic prospects but threaten the livelihoods, identities and cultures of individuals and communities that are inextricably tied to the land. The movement argues that the current trade regime does not respect the 'agrarian citizenship' that food sovereignty demands. This citizenship goes beyond class-based notions of political representation to 'a model of rural action' that protects against the negative impacts of the market as well as state abuses by "encompassing the role of civil society and of democratic communication while also acknowledging ecological limits" (Wittman 2009, p. 808).

There are fears among supporters of the radical movement that if progressive organisations build alliances with reformist institutions within the corporate food regime the 'transformative sense' of food sovereignty may be diluted. Academics and NGOs supporting agroecological alternatives may be drawn into reformist agendas, inadvertently supporting Green Revolution objectives to obtain funding (Holt-Gimènez and Shattuck 2012). Paul Nicolson, member of the Spanish Basque farmers' Union EHNE, cautions against engagement in the non-political programs of progressive movements:

> We must be watchful. We must also be careful, when speaking of food sovereignty, that it is not just limited to access to healthy local products, we must insist on the fact that food sovereignty is intended to bring about change in the economic and social model. There is a big danger, also amongst well-meaning people, as in certain Community Supported Agriculture groups (Nicholson, personal interview, June 2012).

## Conclusion

To reclaim its power as a democratic alternative to the mainstream food system the organic movement needs to join the radical movements in repoliticising the food system through the notion of agrarian citizenship where food is a basic right rather than a commodity. This relies on relocating control of food production and distribution into the hands of producers and consumers rather than corporations and resisting the appropriation of the discourse of food democracy by the dominant actors in the food chain.

If organic farming becomes more intensive and industrialised it may no longer offer a sustainable alternative to industrial agricultural production. Corporate organics may enjoy continued success by drawing on the themes of food democracy, such as eating locally and developing meaningful relationships with producers at the expense of the political project of the progressive movements, particularly those that represent the more transformative values of food sovereignty. The risk is that agroecology may become de-politicised and co-opted by the corporate food regime, as has been the case with organic brands.

Finally, while progressive and generally Western consumers may support and actively promote agroecological alternatives to industrial agriculture through their purchasing behaviour and rhetoric, they are not necessarily actively engaged in the struggle for social justice that requires structural reforms to markets and the redistribution of resources demanded by the radical social movements. While consumption has become a "venue for political action" (Micheletti 2003, p.12) whereby the consumer becomes an agent through 'individualised collective action' exercised in everyday purchasing decisions a deep understanding of the complex issues involved in food politics is required. For "a politically responsible person to know about and respond politically to all those people who daily put breakfast upon our table" he or she must be aware of how "market exchange hides from us the conditions of life of the producers" (Harvey 1996, p.349). An important step is the 'defetishisation' of organics through revelation of the 'real' production relations that lie beneath the corporate marketing campaigns (Johnston et al, 2009, p. 525). This involves interrogating terms such as 'local' and 'small-scale' and interpreting critically the claims that agribusiness corporations makes in regard to food democracy. The devolution of control within the global food system is at the heart of this project.

# References

## Chapter One

Aertsens, J, Verbeke, W, Mondelaers, K, and Huylenbroeck, GV 2009, Personal determinants of organic food consumption: a review, *British Food Journal*, vol. 111, no. 10, pp. 1140-1167.

Agriculture and Consumer Protection 2002, Codex Alimentarius – Organically Produced Foods, viewed 22 April 2013 <http://www.fao.org/DOCREP/005-/Y2772E/Y2772E00.HTM>

Arvola, A, Vassallo, M, Dean, M, Lampila, P, Saba, A, Lähteenmäki, L, et al., 2008, Predicting intentions to purchase organic food: The role of affective and moral attitudes in the Theory of Planned Behaviour, *Appetite*, vol. 50, nos. 2-3, pp. 443-454.

Aschemann, J, Hamm, U, Naspetti, S, and Zanoli, R 2007, The organic market. In *Organic farming: An international history*, ed. W. Lockeretz, 123-51. Wallingford, England: CABI.

BFA 2013, Organic Australia 2013, (formerly Biological farmers of Australia), 20 Reasons to Buy Organics, viewed 24 April 2013 <http://www.bfa-.com.au/WhyOrganics/BenefitsofOrganics-.aspx>.

Bhaskaran, S, Polonsky, M, Gary, J, and Fernandez, S 2006, Environmentally sustainable food production and marketing: opportunity or hype?, *British Food Journal*, vol. 108, no. 8, pp. 677-690.

Bonti-Ankomah, S, and Yiridoe, EK 2006, *Organic and conventional food: a literature review of the economics of consumer perceptions and preference*.Nova Scotia Agriculture College, Organic Agriculture Centre of Canada.

Brennan, C, Gallagher, K, and McEachern, M 2003, A review of the 'consumer interest' in organic meat, *International Journal of Consumer Studies* vol. 27, no. 5, pp. 381-394.

Chakrabarti, S, and Baisya, RK 2007, Purchase motivations and attitudes of organic food buyers, *Decision* (0304-0941),vol. 34, no. 1, pp. 1-22.

Chang, HS, and Zepeda, L 2005, Consumer perceptions and demand for organic food in Australia: focus group discussions, *Renewable Agriculture and Food Systems*, vol. 20, no. 3, pp. 155-167.

Chang, HS, Zepeda, L, and Griffith, G 2005, The Australian organic food products market: overview, issues and research needs, *Australian Agribusiness Review*, vol. 13, no. 5, pp. 1-13

Cicia, G, Giudice, TD, and Scarpa, R 2002, Consumers' perception of quality in organic food: a random utility model under preference heterogeneity and choice correlation from rank-orderings, *British Food Journal*, vol. 104, nos. 3-5, pp. 200-1.

Dangour, AD, Lock, K, Hayter, A, Aikenhead, A, Allen, E, and Uauy, R 2010. Nutrition-related health effects of organic foods: a systematic review, *The American journal of clinical nutrition*, vol. 92, no. 1, pp. 203-210.

Daugbjerg, C, and Halpin, D 2008, Sharpening up research on organics: why we need to integrate sectoral policy research into mainstream policy analysis, *Policy Studies*, vol. 29, no. 4, pp. 393-404.

Dreezens E, Martijn C, Tenbult P, Kok G, Vries N 2005, Food and values: an examination of values underlying attitudes toward genetically modified and organically grown food products. *Appetite*, vol. 44, pp. 115–22.

Essoussi, LH, and Zahaf, M 2008, Decision making process of community organic food consumers: an exploratory study, *Journal of Consumer Marketing*, vol. 25, no. 2, pp. 95-104.

Fearne, A 2008, Organic fruit and vegetables–who buys what and why ... and do we have a clue, *The dunnhumby Academy of Consumer Research, Kent Business School, University of Kent*, viewed 23 April 2013 <http://www.refresh.eu/downloads/conference_08_presentations/Refresh_08_Organics_report.pdf>

Food and Agriculture Organization of the United Nations 2013, viewed 16 April 2013, http://www.fao.org/home/en/

Gil, JM, Gracia, A, and Sanchez, M 2000, Market segmentation and willingness to pay for organic products in Spain, *The International Food and Agribusiness Management Review*, vol. 3, no. 2, pp. 207-226.

Gil, JM, and Soler, F 2006, Knowledge and willingness to pay for organic food in Spain: Evidence from experimental auctions, *Acta Agriculturae Scandinavia Section C*, vol. 3, no. 3-4, pp. 109-124.

Global Industry Analysts 2006, Global Strategic Business Report, viewed 12 May 2013, http://www.strategyr.com/Adipic_Acid_Market_Report.asp.

Green Earth Organics 2008, viewed 14 May 2013, http://www.greenearthorganics.com/

Guido, G, Prete, M, Peluso, A, Maloumby-baka, R, and Buffa, C 2010, The role of ethics and product personality in the intention to purchase organic food products: a structural equation modeling approach, *International Review of Economics*, vol. 57, no. 1, pp. 79-102.

Hall, SB 2011, Australia's organic trilemma: public versus private organic food standardisation, Research Master thesis, University of Tasmania, viewed 23 April 2013, http://eprints.utas.edu.au/12460/

Halpin, D 2004, *The Australian organic industry-a profile*. Canberra: Australia Government Department of Agriculture, Fisheries and Forestry, viewed 25 April 2013, http://www.cabdirect.org/abstracts/20053057396.html;jsessionid=70DE7C237BA554110D787A2A7780CF62.

Halpin, D, and Daugbjerg, C 2008, Associative deadlocks and transformative capacity: engaging in Australian organic farm industry development, *Australian journal of political science*, vol. 43, no. 2, pp. 189-206.

Heckman, J 2005, A history of organic farming: transitions from Sir Albert Howard's war in the soil to USDA national organic program, *Renewable Agriculture and Food Systems*, vol. 21, no. 3, pp. 143-150.

Honkanen, P, Verplanken, B, Olsen, SO 2006, Ethical values and motives driving organic food choice, Journal of Consumer Behaviour, vol. 5 no. 5 pp. 420-30.

Howard, A 1943, *An agricultural testament*. Oxford University Press, London.

Hughner, RS, McDonagh, P, Prothero, A, Shultz Ii, CJ, and Stanton, J 2007. Who are organic food consumers? A compilation and review of why people purchase organic food, *Journal of Consumer Behaviour*, vol. 6, no. 2-3, pp. 94-110.

Hutchins, RK, and Greenhalgh, LA 1997. Organic confusion: sustaining competitive advantage, *Nutrition and Food Science*, vol. 95, no. 6, pp. 11.

IFOAM 2011, About the International Federation of Organic Agriculture Movements (IFOAM) viewed 22 June 2013, http://www.ifoam.org/about_ifoam/index.html

Kearney, J, 2010, Food consumption trends and drivers, *Philosophical Transactions of the Royal Society B: Biological Sciences*, vol.365, no.1554, pp. 2793-2807.

Keating, M, 2010, Organic Agriculture: Its Origins, and Evolution Over Time, viewed 16 June 2013, http://www.huffingtonpost.com/rebecca-gerendasy/organic-agriculture-its-o_b_710722.html

Kihlberg, I, and Risvik, E 2007, Consumers of organic foods—value segments and liking of bread, *Food quality and preference*, vol.18, no.3, pp. 471-481.

Kluger, J, 2010, What's so great about organic food?, *Time Magazine*, vol. 196, no. 10, pp. 1-7.

Krystallis, A, Fotopoulos, C, and Zotos, Y 2006, Organic consumers' profile and their willingness to pay (WTP) for selected organic food products in Greece, *Journal of International Consumer Marketing*, vol. 19, no. 1, pp. 81-106.

Lea, E, and Worsley, T, 2005, Australians' organic food beliefs, demographics and values, *British Food Journal*, vol. 107, no. 11, pp. 855-869.

Lex Column 2013, Spread of GM crops grows Monsanto profit, taken from the Financial Times insert in the Australian Financial Review, 5 April 2013, Fairfax Press, Sydney.

Lockie, S 2006, *Going organic: Mobilizing networks for environmentally responsible food production*. CABI Publishing, London.

Lockie, S, Halpin, D, and Pearson, D, 2006, Understanding the market for organic food. In P. Kristiansen, A. Taji and J. Reganold (Eds.), *Organic agriculture: a global perspective*: CSIRO Publishing.

Lockie, S, Lyons, K, Lawrence, G, and Mummery, K. 2002, Eating 'green': motivations behind organic food consumption in Australia, *Sociologia Ruralis*, vol. 42, no. 1, pp. 23-40.

Loughnan D 2012, *Food shock*, Exisle Publishing Pty Ltd, Wollombi.

Magkos, F, Arvaniti, F, and Zampelas, A 2006, Organic food: buying more safety or just peace of mind? A critical review of the literature, *Critical Reviews in Food Science and Nutrition*, vol. 46, no. 1, pp. 23-56.

Magnusson, M 2004, *Consumer perception of organic and genetically modified foods* PhD Thesis, Uppsala University, viewed 23 June 2013, http://uu.diva-portal.org/smash/record.jsf?pid=diva2:164405

Magnusson, MK, Arvola, A, Hursti, U-KK, Åberg, L, and Sjödén, P-O 2003, Choice of organic foods is related to perceived consequences for human health and to environmentally friendly behaviour, *Appetite*, vol. 40, no. 2, pp. 109-117.

Makatouni, A 2002, What motivates consumers to buy organic food in the UK? Results from a qualitative study, *British Food Journal*, vol. 104, no. 3-5, pp. 345.

Mitchell, A, Kristiansen, P, Bez, N, and Monk, A 2010, *Australian Organic Market Report 2010*. BFA Publication, no. 10/01. Biological Farmers of Australia, Chermside.

Oates, L, Cohen, M, and Braun, L 2012, Characteristics and consumption patterns of Australian organic consumers, *Journal of the Science of Food and Agriculture*, vol. 92, pp. 2782-2787.

OTA 2011, Six myths busted by organic in 2011, Organic Trade Association, viewed 25 June 2013, http://www.organicnewsroom.com/2011/12/six_myths_busted_by_organic_in_1.html

Pieniak, Z, Aertsens, J, and Verbeke, W 2010, Subjective and objective knowledge as determinants of organic vegetables consumption, *Food quality and preference*, vol. 21, no. 6. pp. 581-588.

Pivato, S, Misani, N, and Tencati, A 2008, The impact of corporate social responsibility on consumer trust: the case of organic food, *Business Ethics: A European Review*, vol.17, no.1, pp. 3-12.

Poulston, J, and Yiu, AYK 2011, Profit or principles: Why do restaurants serve organic food?, *International Journal of Hospitality Management*, vol. 30, no. 1, pp. 184-191.

Radman, M 2005, Consumer consumption and perception of organic products in Croatia, *British Food Journal*, vol. 107, no. 4, pp. 263-273.

Roitner-Schobesberger, B, Darnhofer, I, Somsook, S, and Vogl, CR 2008, Consumer perceptions of organic foods in Bangkok, Thailand, *Food Policy*, vol. 33, no. 2, pp. 112-121.

Sahota, A 2009, The global market for organic food and drink. In H. Willer and L. Kilcher (Eds.), *The World of Organic Agriculture*. Bonn, Frick and Geneva. ITC/FIBL/IFOAM.

Schaack, D, and Willer, H 2010, Development of the organic market in Europe, *The World of Organic Agriculture–Statistics and Emerging Trends*, pp. 141-144.

Thøgersen, J 2011, Green shopping: For selfish reasons or the common good, *American Behavioral Scientist*, vol. 55, no. 8, pp. 1052-1076.

Thøgersen, J 2009, Consumer decision-making with regard to organic food products. In Vaz, P Nijkamp and J-L Rastoin (Eds.), *Traditional food production and rural sustainable development: an European challenge*, Ashgate Publishing Ltd, Farnham.

Thøgersen, J 2010, Country differences in sustainable consumption: The case of organic food, *Journal of Macromarketing*, vol. 30, no. 2, pp. 171-185.

Torjusen, H, Lieblein, G, Wandel, M, and Francis, CA 2001, Food system orientation and quality perception among consumers and producers of organic food in Hedmark County, Norway, *Food Quality and Preference*, vol. 12, no. 3, pp. 207-216.

United States Department of Agriculture 2005, Regional coverage: Beijing viewed 12 April 2013 http://www.usdachina.org/en_pop.asp?id=66[14.

Walley, KE., Custance PR., and Parsons ST 2009, Controversies in Food and Agricultural Marketing: The Consumer's View. In A. Lindgreen, MK Hingley, and J Vanhamme (Eds), *The Crisis of Food Brands. Sustaining safe, Innovative and Competitive Food Supply*, pp. 197-219, Gower Publishing Limited. Burlington.

Yussefi, M 2008, *The World of Organic Agriculture: Statistics and Emerging Trends 2008*. Willer, H and Sorensen, N (Eds.). Earthscan.

Zanoli, R, and Naspetti, S 2002, Consumer motivations in the purchase of organic food, *British Food Journal*, vol. 104, pp. 643.

Zhao, X, Chambers, E, Matta, Z, Loughin, T, and Carey, E 2007, Consumer sensory analysis of organically and conventionally grown vegetables, *Journal of Food Science*, vol. 72, no. 2, pp. 87-91.

## Chapter Two

Bellon, S, and Lamine, C 2009, Conversion to Organic Farming: A Multidimensional Research Object at the Crossroads of Agricultural and Social Sciences – A Review, *Sustainable Agriculture*, vol. 77. no. 112, pp. 653-672.

Benbrook, C, Davis, DR, and Andrews, PK 2009, Methodologic flaws in selecting studies and comparing nutrient concentrations led Dangour et al. to miss the emerging forest amid the trees, *The American journal of clinical nutrition*, vol. 90, no. 6, pp. 1700-1701.

Benbrook, C, Zhao, X, Yáñez, J, Davies, N, and Andrews, P 2008, New evidence confirms the nutritional superiority of plant-based organic foods, *The Organic Center: Foster, RI*, viewed 24 June 2013, https://www.organic-center.org/tocpdfs/NutrientContentExecSummary.pdf

Bengtsson, J, Ahnström, J, and Weibull, ANNC 2005, The effects of organic agriculture on biodiversity and abundance: a meta-analysis, *Journal of applied ecology*, vol. 42, no. 2, pp. 261-269.

Blake D 2012, 13 Benefits of Organic Food, a report for USDA, viewed 23 February 2013 http://ecoscraps.com/13-benefits-organic-food/

Brandt, K, and Mølgaard, JP 2001. Organic agriculture: does it enhance or reduce the nutritional value of plant foods?, *Journal of the Science of Food and Agriculture*, vol. 81, no. 9, pp. 924-931.

Dangour, AD, Lock, K, Hayter, A, Aikenhead, A, Allen, E, and Uauy, R 2009, Nutrition-related health effects of organic foods: a systematic review, *The American journal of clinical nutrition*, vol. 90, pp. 680-685.

Dangour, AD, Lock, K, Hayter, A, Aikenhead, A, Allen, E, and Uauy, R 2010, Nutrition-related health effects of organic foods: a systematic review, *The American journal of clinical nutrition*, vol. 92, no. 1, pp. 203-210.

Flessa, H, Ruser, R, Dörsch, P, Kamp, T, Jimenez, MA, Munch, JC, et al. 2002 Integrated evaluation of greenhouse gas emissions ($CO_2$, $CH_4$, $N_2O$) from two farming systems in southern Germany, *Agriculture, Ecosystems and Environment*, vol. 91, nos. 1-3, pp. 175-189.

Fliessbach, A, Oberholzer, HR, Gunst, L, and Mäder, P 2007, Soil organic matter and biological soil quality indicators after 21 years of organic and conventional farming, *Agriculture, Ecosystems and Environment*, vol. 118, no. 1, pp. 273-284.

Gibney M 2012, No scientific evidence showing organic is better, *The Irish Times*, Thursday, 5 July 2012, viewed 22 July 2012, http://www.irishtimes.com/newspaper/opinion-/2012/0705/1224319427071.html

Grandy, AS, and Robertson, GP 2007, Land-use intensity effects on soil organic carbon accumulation rates and mechanisms, *Ecosystems*, vol. 10, no. 1, pp. 59-74.

Griffiths, MA, Cook, MD, Eggett, LD, and Christensen, JM 2011, A retail market study of organic and conventional potatoes (Solanum tuberosum): mineral content and nutritional implications, *International Journal of Food Sciences and Nutrition*, vol. 63, no. 4, pp. 393-401.

Harper, GC, and Makatouni, A 2002, Consumer perception of organic food production and farm animal welfare, *British Food Journal*, vol. 104, no. 3, pp. 287-299.

Hill, H, and Lynchehaun, F 2002, Organic milk: attitudes and consumption patterns, *British Food Journal*, vol. 104, no. 7, pp. 526-542.

Hole, DG, Perkins, AJ, Wilson, JD, Alexander, IH, Grice, PV, and Evans, AD 2005, Does organic farming benefit biodiversity?, *Biological conservation*, vol. 122, no. 1, pp. 113-130.

Hope J 2012, Organic food isn't healthier and no safer than produce grown with pesticides, finds biggest study of its kind, viewed 3 September 2012, http://www.dailymail.co.uk/health/article-2197854/Organic-better-produce-grown-pesticides-say-Stanford-University-scientists.html#ixzz2NfWuOU6I

Huber, M, Bakker, MH, Dijk, W, Prins, HAB, and Wiegant, FAC 2012, The challenge of evaluating health effects of organic food; operationalisation of a dynamic concept of health, *Journal of the Science of Food and Agriculture*, vol. 92, no. 14, pp. 2766-2773.

Huber, M, Rembialkowska, E, Srednicka, D, Bügel, S, and van de Vijver, LPL 2011, Organic food and impact on human health: Assessing the status quo and prospects of research, *NJAS-Wageningen Journal of Life Sciences*, vol. 58, no. 3, pp. 103-109.

Hunter, D, Foster, M, McArthur, JO, Ojha, R, Petocz, P, and Samman, S 2011, Evaluation of the micronutrient composition of plant foods produced by organic and conventional agricultural methods, *Critical reviews in food science and nutrition*, vol. 51, no. 6, pp. 571-582.

IFOAM 2006, *International federation of organic agriculture movements annual reports 2005 and 2006*. viewed 3 January 2012, http://www.ifoam.org/sites/default/files/page/files/ifoam_annual_report_2005-2006.pdf

Jensen, MM, Jørgensen, H, Halekoh, U, Watzl, B, Thorup-Kristensen, K, and Lauridsen, C 2012, Health biomarkers in a rat model after intake of organically grown carrots, *Journal of the Science of Food and Agriculture*, vol. 92, no. 15, pp. 2936-2943.

Kahl, J, Baars, T, Bügel, S, Busscher, N, Huber, M, Kusche, D, et al 2011, Organic food quality: a framework for concept, definition and evaluation from the European perspective, *Journal of the Science of Food and Agriculture,* vol. 92, no. 14, pp. 2760-2765.

Kouba, M 2003, Quality of organic animal products, *Livestock Production Science,* vol. 80, nos. 1-2, pp. 33-40.

Kristensen, M, Østergaard, LF, Halekoh, U, Jørgensen, H, Lauridsen, C, Brandt, K, et al. 2008, Effect of plant cultivation methods on content of major and trace elements in foodstuffs and retention in rats, *Journal of the Science of Food and Agriculture,* vol. 88, no. 12, pp. 2161-2172.

Lairon, D 2011, Nutritional quality and safety of organic food, *Sustainable Agriculture,* vol. 2, pp. 99-110.

Loughnan D 2012, *Food shock*, Exisle Publishing, NSW.

Mäder, P, Fliessbach, A, Dubois, D, Gunst, L, Fried, P, and Niggli, U 2002, Soil fertility and biodiversity in organic farming, *Science,* vol. 296, no. 5573, pp. 1694.

Magkos, F, Arvaniti, F, and Zampelas, A 2003, Putting the safety of organic food into perspective, *Nutrition Research Reviews,* 16, pp. 211.

Mondelaers, K, Aertsens, J, and Van Huylenbroeck, G 2009, A meta-analysis of the differences in environmental impacts between organic and conventional farming, *British Food Journal,* vol.111, no. 10, pp. 1098-1119.

OTA 2011, Six myths busted by organic in 2011, Organic Trade Association, viewed 11 January 2012, http://www.organicnewsroom.com/2011/12/six_myths_busted_by_organic_in_1.html

Palmer, C 2012, Organic food no better for you – study: The Conversation, 4 September 2012, Melbourne, viewed 22 October 2013, http://theconversation.edu.au/organic-food-no-better-for-you-study-9300

Palupi, E, Jayanegara, A, Ploeger, A, and Kahl, J 2012, Comparison of nutritional quality between conventional and organic dairy products: a meta-analysis, *Journal of the Science of Food and Agriculture*, vol. 92, no.14, pp. 2774-2781.

Reganold, JP, Andrews, PK, Reeve, JR, Carpenter-Boggs, L, Schadt, CW, Alldredge, JR, et al 2010, Fruit and soil quality of organic and conventional strawberry agroecosystems, *PloS one*, vol. 5, no. 9, e12346.

Rosen, JD 2010, A review of the nutrition claims made by proponents of organic food, *Comprehensive Reviews in Food Science and Food Safety*, vol. 9, no. 3, pp. 270-277.

Rundlöf, M, and Smith, HG 2006, The effect of organic farming on butterfly diversity depends on landscape context, *Journal of applied ecology*, vol. 43, no. 6, pp. 1121-1127.

Sargeant, JM, Rajic, A, Read, S, and Ohlsson, A 2006, The process of systematic review and its application in agri-food public-health, *Preventive veterinary medicine*, vol. 75, no. 3, pp. 141-151.

Søltoft, M, Bysted, A, Madsen, KH, Mark, AB, Bügel, SG, Nielsen, J, et al 2011, Effects of organic and conventional growth systems on the content of carotenoids in carrot roots, and on intake and plasma status of carotenoids in humans, *Journal of the Science of Food and Agriculture*, vol. 91, no. 4, pp. 767-775.

The Soil Association 2009, Soil Association response to the FSA's Organic Review. *Soil Association Producer eNews*, viewed 14 April 2013, http://www.soilassociation.org/LinkClick.aspx?fileticket=G8qHwFuy5VQpercent3D

Torjusen, H, Brantsæter, AL, Haugen, M, Lieblein, G, Stigum, H, Roos, G, et al. 2010, Characteristics associated with organic food consumption during pregnancy; data from a large cohort of pregnant women in Norway, *BMC public health*, vol. 10, no. 1, pp. 775.

Tscharntke, T, Klein, AM, Kruess, A, Steffan-Dewenter, I, and Thies, C 2005, Landscape perspectives on agricultural intensification and biodiversity-ecosystem service management, *Ecology letters*, vol. 8, no. 8, pp. 857-874.

Van de Vijver, LPL, and van Vliet, MET 2012, Health effects of an organic diet-consumer experiences in the Netherlands, *Journal of the Science of Food and Agriculture*, vol. 92, no. 14, pp. 2923-2927.

Williams, P, and Hammitt, J 2001, Perceived Risks of Conventional and Organic Produce: Pesticides, Pathogens, and Natural Toxins, *Risk Analysis*, vol. 21, no. 2, pp. 319-330.

Woese, K, Lange, D, Boess, C, and Bögl, KW 1997, A comparison of organically and conventionally grown foods – results of a review of the relevant literature, *Journal of the Science of Food and Agriculture*, vol. 74, no. 3, pp. 281-293.

Young, I, Rajic, A, Wilhelm, BJ, Waddell, L, Parker, S, and McEwen, SA 2009, Comparison of the prevalence of bacterial enteropathogens, potentially zoonotic bacteria and bacterial resistance to antimicrobials in organic and conventional poultry, swine and beef production: a systematic review and meta-analysis, *Epidemiology and infection*, vol. 137, no. 9, pp. 1217-1232.

## Chapter Three

AIF 1916, Australian Imperial Force, Attestation Paper of Persons Enlisted for Service Abroad, No. 5362, GENONI, ERNESTO, 25/2/16. Canberra: National Archives of Australia.

Balfour, E B 1959a, *Letter from Australia*, No. 3. Mother Earth, vol. 10, no. 8, pp. 701-722.

Balfour, E B 1959b, *Letter from Australia*, No. 4. Mother Earth, vol. 11, no. 1, pp. 21-48.

Balfour, E B 1960, *Letter from Australia*, No. 5. Mother Earth, vol. 11, no. 4, pp. 397-421.

Bayles, R J 1947, Letter to James Melrose, Esq., Barton, Campbell Town from LSAT. Unpublished MS, 2 April, 2 pp., NS248/1/1-4; Hobart: Archives Office of Tasmania.

Berry, V 2012, Rigby Instant Books, viewed 11 April 2013, http://vanessaberryworld.wordpress.com/2012/01/22/rigby-instant-books/.

Griffin, MM 1949, *The Magic of America* MS typescript, The Art Institute of Chicago, Ryerson and Burnham Libraries (Archives), Chicago, IL; www.artic.edu.

Hills, LD 1989, *Fighting Like the Flowers: An Autobiography*. Hartland Bideford, Devon: Green Books.

Jones, R 2010, *Green Harvest: A History of Organic Farming and Gardening in Australia*, CSIRO Publishing, Melbourne.

LSAT 1947, *Rules and Constitution, The Living Soil Association of Tasmania*. Unpublished manuscript, Hobart.

Martin, SK 1975, Application for Incorporation of an Association – The Organic Food Movement. Signed 26 May 1975; Incorporated 9 June 1975. Adelaide: Office of the Registrar of Companies.

McKell, WJ 1946, Soil – the basis of our civilisation. *Organic Farming Digest*, vol. 1, no. 1, pp. 1-2.

Northbourne, L 1940, *Look to the Land*. London: Dent.

OFA 2013, Welcome to the Organic Federation of Australia, viewed 5 April 2013: www.ofa.org.au.

OIECC 2009, National Standard for Organic and Bio-Dynamic Produce, Edition 3.4, Last updated 1 July 2009. Canberra: Organic Industry Export Consultative Committee (OIECC).

Paull, J 2008, The lost history of organic farming in Australia. *Journal of Organic Systems*, vol. 3, no. 2, pp. 2-17.

Paull, J 2009, The Living Soil Association: Pioneering organic farming and innovating social inclusion. *Journal of Organic Systems*, vol. 4, no. 1, pp. 15-33.

Paull, J 2010, From France to the World: The International Federation of Organic Agriculture Movements (IFOAM). *Journal of Social Research and Policy*, vol. 1, no. 2, pp. 93-102.

Paull, J 2011a, Attending the First Organic Agriculture Course: Rudolf Steiner's Agriculture Course at Koberwitz, 1924. *European Journal of Social Sciences*, vol. 21, no. 1, pp. 64-70.

Paull, J 2011b, The Betteshanger Summer School: Missing link between biodynamic agriculture and organic farming. *Journal of Organic Systems*, vol. 6, no. 2, pp. 13-26.

Paull, J 2011c, Biodynamic Agriculture: The journey from Koberwitz to the World, 1924-1938. *Journal of Organic Systems*, vol. 6, no. 1, pp. 27-41.

Paull, J 2011d, Organics Olympiad 2011: Global Indices of Leadership in Organic Agriculture. *Journal of Social and Development Sciences*, vol. 1, no. 4, pp. 144-150.

Paull, J 2011e, The secrets of Koberwitz: The diffusion of Rudolf Steiner's Agriculture Course and the founding of Biodynamic Agriculture. *Journal of Social Research and Policy*, vol. 2, no. 1, pp. 19-29.

Paull, J 2011f, The Soil Association and Australia: From Mother Earth to Eve Balfour. *Mother Earth*, vol. 4 (Spring), pp. 13-17.

Paull, J, and Hennig, B 2013, The World of Organic Agriculture–Density-equalizing Map. *The World of Organic Agriculture: Statistics and Emerging Trends 2013*, viewed 11 April 2013, http://orgprints.org/22490/.

Paull, J, 2013, The Rachel Carson letters and the making of Silent Spring. *Sage Open*, 3 (July-September), 1-12.

Pfeiffer, E 1938, *Bio-Dynamic Farming and Gardening: Soil Fertility Renewal and Preservation* (F. Heckel, Trans.), Anthroposophic Press. New York.

Pike, D, 1967, *Paradise of Dissent: South Australia 1829-1857*, Vol. 2, Melbourne University Press, Melbourne.

Reed, MJ 2010, *Rebels for the Soil: The Rise of the Global Food and Farming Movement*. London, Earthscan.

Rodale, JI 1946, Is our health related to the soil? *Organic Farming Digest*, vol. 1, no. 1, pp. 22-27.

Rodale, JI 1942, *Organic Farming and Gardening*. Emmaus, Rodale Press, Pennsylvania,

SA 2009, Australian Standard for Organic and Biodynamic Products, 6000-2009. Sydney: Standards Australia (SA).

Spathopoulos, W 2007, The Crag: Castlecag 1924 -1938. Blackheath, Brandl and Schlesinger, NSW.

Steiner, R 1924a, Agriculture Course (Printed for private circulation only; 1929, first English language edition; George Kaufmann Trans). Dornach, Switzerland, Goetheanum.

Steiner, R 1924b, To All Members: The Meetings at Breslau and Koberwitz; the Waldorf School; the longings of the Youth. *Anthroposophical Movement*, vol. 1, pp. 17-18.

Steiner, R, 1924c, To All Members: The Meetings at Koberwitz and Breslau. *Anthroposophical Movement*, vol. 1, pp. 9-11.

Stevenson, G, 2009, *Ahead of their Time: A History of the Organic Gardening and Farming Society of Tasmania*. Somerset, Dunghill Press, Tasmania.

The Executive Officers 1954, *Farewell. Farm and Garden Digest* (incorporating *Organic Farming Digest*), vol. 3, no. 5, pp. 1-3.

White, HF, and Hicks, CS 1953, *Life from the Soil*, Longmans Green and Co, Melbourne.

Willer, H, and Kilcher, L 2011, The World of Organic Agriculture: Statistics and Emerging Trends 2011: Bonn: International Federation of Organic Agriculture Movements (IFOAM). *Frick, Switzerland: Research Institute of Organic Agriculture (FiBL)*.

Willer, H and Kilcher, L 2012, *The World of Organic Agriculture: Statistics and Emerging Trends 2012*: Bonn: International Federation of Organic Agriculture Movements (IFOAM); Frick, Switzerland: Research Institute of Organic Agriculture (FiBL).

Willer, H, Lernoud, J, and Kilcher, L 2013, *The World of Organic Agriculture: Statistics and Emerging Trends 2013*: Frick, Switzerland: Research Institute of Organ Agriculture (FiBL) and Bonn: International Federation of Organic Agriculture Movements (IFOAM).

Willer, H, and Yussefi, M, 2000, *Organic Agriculture World-Wide: Statistics and Perspectives*. Bad Durkheim, Germany: Stiftung Ökologie and Landbau (SÖL).

Williams, Louise (attrib.). (c. 1984). *Robert Henry Williams*: (pp. 3; unpublished typescript MS; private collection).

Windram, A 1972, *Organic Food Movement Newsletter*, December.

Windram, A, 1973, Recommendations from the Standards Committee. *Bulletin of the Organic Food Movement*. A section of the Soil Assoc. (SA Branch).

Windram, A 1975, *Organic Gardening*. Adelaide: Rigby.

Wynen, E 2007, Organic Farming Australia. In H. Willer and M. Yussefi (Eds.). *The World of Organic Agriculture, Statistics and Emerging Trends*. Bonn: International Federation of Organic Agriculture Movements (IFOAM).

## Chapter Four

ABS (Australian Bureau of Statistics), 2011, Household Expenditure Survey, Australia: Detailed Expenditure Items, 2003-04 (Reissue) (Cat. No. 6535.0.55.001), ABS, Melbourne.

BFA (Biological Farmers of Australia Ltd), 2008, *Australian Organic Market Report 2008*, BFA, Chermside, Queensland.

DAFF (Department of Agriculture, Fisheries and Forestry), 2011a, Australian food statistics 2010-11, viewed 12 July 2013, http://www.daff.gov.au/agriculture-food/food/publications/afs.

DAFF (Department of Agriculture, Fisheries and Forestry), 2011b, Agricultural commodity statistics 2011, viewed 12 July 2013, http://www.daff.gov.au/agriculture-food/food/publications/afs/australian-food-statistics.

Euromonitor, 2010, *Health and Wellness Statistics*, viewed 11 April 2013, euromonitor.com/health-and-wellness-tourism.

Euromonitor, 2011a, *Certified organic: recession no serious threat to organic food and drinks* (Part 1), viewed 11 April 2013, euromonitor.com/certified-organic-recession-no-serious-threat-to-organic-food-and-drinks-part-1/report.

Euromonitor 2011b, *Organic Market Size Statistics*, viewed 11 April 2013, portal.euromonitor.com/Portal/Pages/Search/SearchResultsList.aspx.

Euromonitor 2012a, *Health and Wellness Statistics*, viewed 11 April 2013, euromonitor.com/health-and-wellness-tourism.

Euromonitor 2012b, *Certified organic: winning strategies of key organic players* (Part 2), viewed 11 April 2013, euromonitor.com/certified-organic-winning-strategies-of-key-organic-players-part-2/report.

Halpin, D 2004, A farm-level view of the Australian organic industry, in Halpin, D (ed.), *The Australian Organic Industry: A Profile*, Department of Agriculture, Fisheries and Forestry, Canberra, pp. 1-29.

Mitchell, A, Kristiansen, P, Bez, N, and Monk, A, 2010, *Australian Organic Market Report 2010*, BFA Ltd, Chermside, Brisbane.

Monk, A, Mascitelli, B, Lobo, A, Chen, J and Bez, N 2012, *Australian Organic Market Report 2012*, BFA Ltd, Chermside, Brisbane.

UK Soil Association, 2011, *Organic Market Report 2011*, Soil Association, viewed 22 March 2012, soilassociation.org/marketreport.

Willer, H, and Kilcher, L, 2012, The world of organic agriculture: statistics and emerging trends 2012, *IFOAM*, Bonn, and FiBL, Frick, Switzerland.

## Chapter Five

Adams, DC, and Salois, MJ 2010, Local versus organic: A turn in consumer preferences and willingness-to-pay. *Renewable Agriculture and Food Systems*, vol. 25, no. 4, pp. 331-341.

Anders, S, and Moeser, A, 2008, Assessing the demand for value-based organic meats in Canada: a combined retail and household scanner-data approach. *International Journal of Consumer Studies*, vol. 32, no. 5, pp. 457-469.

Batte, MT, Hooker, NH, Haab, TC, and Beaverson, J 2007, Putting their money where their mouths are: Consumer willingness to pay for multi-ingredient, processed organic food products. *Food Policy*, vol. 32, no. 2, pp. 145-159.

Bellows, AC, Onyango, B, Hallman, WK, and Diamond, A 2008, Understanding consumer interest in organics: production values vs. purchasing behavior. *Journal of Agricultural and Food Industrial Organisation*, vol. 6, no. 1, pp. 1-31.

Bonti-Ankomah, S, and Yiridoe, EK 2006, *Organic and conventional food: a literature review of the economics of consumer perceptions and preference*: Nova Scotia Agriculture College, Organic Agriculture Centre of Canada.

Chinnici, G, D'Amico, M, and Peorino, B 2002, A multivaritate statistical analysis on the consumers of organic products *British Food Journal*, vol. 104, no. 3/4/5, pp. 187-199.

Chryssochoidis, G, Krystallis, A, and Perreas, P 2007, Ethnocentric beliefs and country-of-origin (COO) effect: Impact of country, product and product attributes on Greek consumers' evaluation of food products. *European Journal of Marketing*, vol. 41, no. 11/12, pp. 1518-1544.

Dangour, AD, Lock, K, Hayter, A, Aikenhead, A, Allen, E, and Uauy, R 2010. Nutrition-related health effects of organic foods: a systematic review. *The American journal of clinical nutrition*, vol. 92, no. 1, pp. 203-210.

Everage, L 2002, Understanding the LOHAS lifestyle. *Gourmet Retailer*, vol. 23, no. 10, pp. 82-87.

Gil, JM, Gracia, A, and Sanchez, M 2000, Market segmentation and willingness to pay for organic products in Spain. *The International Food and Agribusiness Management Review*, vol. 3, no. 2, pp. 207-226.

Henryks, J, and Pearson, D 2010, Marketing communications create confusion: Perception versus reality for Australian organic food consumers. In *Australian and New Zealand Communications Association Conference: Media Democracy and Change*, pp. 7-9.

Hjelmar, U 2011, Consumers' purchase of organic food products. A matter of convenience and reflexive practices. *Appetite*, vol. 56, no. 2, pp. 336-344.

Hughner, RS, McDonagh, P, Prothero, A, Shultz Ii, CJ, and Stanton, J 2007, Who are organic food consumers? A compilation and review of why people purchase organic food. *Journal of Consumer Behaviour*, vol. 6, nos. 2-3, pp. 94-110.

Janssen, M, and Hamm, U, 2011, Consumer perception of different organic certification schemes in five European countries. *Organic Agriculture*, vol. 1, no. 1, pp. 31-43.

Janssen, M, Heid, A, and Hamm, U 2010, Is there a promising market in between organic and conventional food? Analysis of consumer preferences. *Renewable Agriculture and Food Systems*, vol. 24, no. 3, p. 205.

Krystallis, A, and Chryssohoidis, G 2005, Consumers' willingness to pay for organic food: factors that affect it and variation per organic product type. *British Food Journal*, vol. 107, no. 5, pp. 320-343.

Krystallis, A, Fotopoulos, C, and Zotos, Y 2006, Organic consumers' profile and their willingness to pay (WTP) for selected organic food products in Greece. *Journal of International Consumer Marketing*, vol. 19, no. 1, pp. 81-106.

Kuhar, A, and Juvancic, L 2010, Determinants of purchasing behaviour for organic and integrated fruits and vegetables in Slovenia. *Agricultural Economics Review*, vol. 11, no. 2, pp. 70-83.

Kumar, V, Jones, E, Venkatesan, R, and Leone, RP 2011, Is Market Orientation a Source of Sustainable Competitive Advantage or Simply the Cost of Competing? *Journal of Marketing*, vol.75, no. 1, pp. 16-30.

Lin, BH, Smith, TA, and Huang, CL 2008, Organic premiums of US fresh produce. *Renewable Agriculture and Food Systems (formerly American Journal of Alternative Agriculture)*, vol. 23, no. 3, pp. 208-216.

Lobo, A, Mascitelli, B, and Chen, J 2012, *Investigating Consumers' Purchase Behaviour of Organic Food in Australia*. Paper presented at the GBATA, 2012, New York.

Lobo, A, Mascitelli, B, and Chen, J 2012, Investigating Consumers' Purchase Behaviour of Organic Food Products in Victoria: PhD thesis, Faculty of Business Enterprise, Swinburne University of Technology.

Lockie, S, Halpin, D, and Pearson, D 2006, Understanding the market for organic food In P. Kristiansen, A. Taji and J. Reganold (Eds.), *Organic agriculture: a global perspective*: CSIRO Publishing, Collingwood.

Lockie, S, Lyons, K, Lawrence, G, and Grice, J 2004, Choosing organics: a path analysis of factors underlying the selection of organic food among Australian consumers. *Appetite*, vol. 43, no. 2, pp. 135-146.

Michaelidou, N, and Hassan, LM 2010, Modeling the factors affecting rural consumers' purchase of organic and free-range produce: A case study of consumers' from the Island of Arran in Scotland, UK. *Food Policy*, vol. 35, no. 2, pp. 130-139.

Millock, K, Wier, M, and Andersen, LM 2004, *Consumer demand for organic foods – attitudes, values and purchasing behaviour*. Paper presented at the Thirteenth Annual Conference of European Association of Environmental and Resource Economics Budapest, Hungary, viewed 11 April 2013, http://orgprints.org/4754/

Ngobo, PV 2011, What Drives Household Choice of Organic Products in Grocery Stores? *Journal of Retailing*, vol. 87, no. 1, pp. 90-100.

O' Donovan, P, and McCarthy, M 2002, Irish consumer preference for organic meat. *British Food Journal*, vol. 104, no. 3-5, pp. 353-370.

Soler, F, Gil, JM, and Sánchez, M 2002, Consumers' acceptability of organic food in Spain. *British Food Journal*, vol. 104, nos. 8/9, pp. 670-687.

Thøgersen, J 2010, Country differences in sustainable consumption: The case of organic food. *Journal of Macromarketing*, vol. 30, no. 2, pp. 171-185.

Torjusen, H, Sangstad, L, O'Doherty Jensen, K, and Kjærnes, U 2004, European consumers' conceptions of organic food: A review of available research. National Institute for Consumer Research, SIFO, Oslo, viewed 22 May 2013, http://orgprints.org/2490/

Tsakiridou, E, Mattas, K, and Tzimitra-Kalogianni, I 2006, The influence of consumer characteristics and attitudes on the demand for organic olive oil. *Journal of International Food and Agribusiness Marketing*, vol. 18, nos. 3/4, pp. 23-31.

Ureña, F, Bernabéu, R, and Olmeda, M 2008, Women, men and organic food: differences in their attitudes and willingness to pay: a Spanish case study. *International Journal of Consumer Studies*, vol. 32, no. 1, pp. 18-26.

Wier, M, O'Doherty Jensen, K, Andersen, LM, and Millock, K 2008, The character of demand in mature organic food markets: Great Britain and Denmark compared. *Food Policy*, vol. 33, no. 5, pp. 406-421.

Yang, KCC 2004, A comparison of attitudes towards internet advertising among lifestyle segments in Taiwan. *Journal of Marketing Communications*, vol. 10, no. 3, pp. 195-212.

Zanoli, R, and Naspetti, S 2002, Consumer motivations in the purchase of organic food. *British Food Journal*, vol. 104, no. 8, pp. 643-653.

## Chapter Six

Australian Conservation Foundation 2005, Cut Back on 'Food Miles viewed 30 May 2013 http://www.acfonline.org.au.

Celuch, K., Murphy, GB, and Callaway, SK 2007, More bang for your buck: Small firms and the importance of aligned information technology capabilities and strategic flexibility. *Journal of High Technology Management Research* vol.17, no. 2, pp. 187-197.

Defra 2008, *Understanding of Consumer Attitudes and Actual Purchasing Behaviour, with Reference to Local and Regional Foods*, Department of Food, Rural Affairs, London.

Dimitri, C, and Oberholtzer, L 2009, Meeting market demand in the organic sector: Handler-supplier relationships in the face of tight supply. *Renewable Agriculture and Food Systems*, vol. 24, no. 2, pp. 137-145.

Gaballa, S, and Abraham, AB 2007, Food Miles in Australia: A preliminary study of Melbourne, Victoria. Melbourne, Australia: Centre for Education and Research in Environmental Strategies (CERES)– Community Environment Park.

Khan, ZR, and Pickett, JA, 2010, Push-Pull Strategy for insect Pest Management, http://entomology.ifas.ufl.edu/capinera/eny5236/pest2/content/14/29_push_pull_strategy.pdf Accessed August 17, 2013

Kottila, MR, Maijala, A, and Rönni, P 2005, The organic food supply chain in relation to information management and the interaction between actors. *ISOFAR*, viewed 23 April 2013, http://orgprints.org/4402/

Kujala, J, and Kristensen, H 2005, Organic marketing initiatives and rural development-lessons learned for the organic industry. *NJF Report*, vol. 1 no. 1, pp. 25-29.

Mascitelli, B, Lobo, A, and Chen, J 2011, *Investigating supply chain management practices in the Victorian Organic Fresh Fruit and Vegetable Sector*, PhD thesis, Faculty of Business and Enterprise, Swinburne University of Technology, Hawthorn.

Pearson, D, Henryks, J, and Jones, H 2011, Organic food: What we know (and do not know) about consumers. *Renewable Agriculture and Food Systems (formerly American Journal of Alternative Agriculture)*, vol. 26. no. 2, pp. 171-177.

Perosio, DJ, McLaughlin, EW, and Cuellar, S 2003, *A Menu of Opportunity: Produce in the Foodservice Industry*. Paper presented at the FreshTrack 2003, Produce Marketing Association, Newark, Delaware.

Richter, T 2004, *Marketing organic products via European retail chains*. viewed 11 April 2013, http://orgprints.org/1893/1/richter-2004-biofach-retail.pdf

Stadtler, H 2005, Supply chain management and advanced planning basics, overview and challenges. *European Journal of Operational Research*, vol. 163, no. 3, pp. 575-588.

Sullivan, T 2004, *Marketing and distribution concepts – the view of a supermarket*. Paper presented at the European hearing on organic food and farming – towards a European action plan, Brussels.

Tavella, E, and Hjortsø, CN 2011, *Design and manage local organic food supply chains: Benefits of using soft systems methodology*. Paper presented at the 55th Annual

Meeting of the International Society for the Systems Sciences University of Hull Business School, UK.

The Food Institute, 2005, Natural products shoppers buy more, but shop in fewer places, *The Food Institute Report*.

Tondel, F, and Woods, TA 2006, *Supply Chain Management and the Changing Structure of U.S. Organic Produce*. Paper presented at the American Agricultural Economics Association, 2006 Annual meeting, July 23-26, Long Beach, CA.

UrburnFarmingOz. (2013). Food Miles. Retrieved 24/03/2013, from http://urbanfarmingoz.com.au/about-uf/food-miles.html

## Chapter Seven

Australian Organic Market Report 2010, viewed 11 June 2013, http://www.bfa.com.au/IndustryResources/BFAPublications/AustralianOrganicMarketReport/tabid/123/Default.aspx#10 .

Bachman L, Cruzada E, Wright S, 2009, *Food Security and Farmer Empowerment*, MASIPAG, 2611 Carbern Village, Anos Los Banos, Laguna 4000, Philippines, 2009, ISBN 078-971-94381-0-6.

Badgley, C, Moghtader, J, Quintero, E, Zakem, E, Chappell, MJ, Aviles-Vazquez, K, Samulon, A, and Perfecto, I, 2007, Organic agriculture and the global food supply. *Renewable agriculture and food systems*, vol. 22, no. 2, pp. 86-108.

Bradford JM 2008, Organic Pecans: Another Option for Growers, Agricultural Research magazine. US Agricultural Research Service (ARS).

Cacek T and Langner LL 1986, The economic implications of organic farming, 1986, American Journal of Alternative Agriculture, Vol. 1, No. 1, pp. 25-29.

Chang J Pers Com, 2012, Jennifer Change is the Executive Director of IFOAM Asia and presented this information at a meeting of IFOAM Asia in Manila, Philippines, in October 2012.

Delate, K, Cambardella, C, Chase, C, Johanns, A, and Turnbull, R 2013, The Long-Term Agroecological Research (LTAR) experiment supports organic yields, soil quality, and economic performance in Iowa. Online. *Crop Management*, viewed 22 March 2013, http://www.plantmanagementnetwork.org/pub/cm/symposium/organic/farm/LTAR/

de Ponti T, Rijk B, van Ittersum M 2012,The crop yield gap between organic and conventional agriculture, *Agricultural Systems*, vol. 108, pp 1-9.

Drinkwater LE, Wagoner P and Sarrantonio M 1998, Legume-based cropping systems have reduced carbon and nitrogen losses. *Nature*, vol. 396, pp. 262-265.

Edwards S, Egziabher T and Araya H 2011, Successes and Challenges in Ecological Agriculture: in Experiences from Tigray, Ethiopia, Eds. Lim L.C., Edwards S. and El-Hage Scialabba N., in *Climate Change and Food Systems Resilience in Sub-Saharan Africa*, Food and Agriculture Organization of the United Nations, ISBN 978-92-5-106876-2.

FAO 2011, Save and Grow (Rome, 2011), chapter 1, viewed 30 June 2013, http://www.fao.org/ag/save-and-grow/en/1/index.html

Hassanali, A, Herren, H, Khan, ZR, Pickett, JA, Woodcock, CM 2008, Integrated pest management: the push–pull approach for controlling insect pests and weeds of cereals, and its potential for other agricultural systems including animal husbandry, Philosophical Transactions of the Royal Society B. 363: 611-621.

IFOAM 2013, viewed 30 June 2013, http://www.ifoam.org/en/organic-landmarks/definition-organic-agriculture.

LaSalle T and Hepperly P 2008, Regenerative organic farming: A solution to global warming. The Rodale Institute, USA.

Lotter DW, Seidel R and Liebhart W 2003, The performance of organic and conventional cropping systems in an extreme climate year. *American Journal of Alternative Agriculture*, vol. 18, no. 3, pp. 146-154.

Newspoll, 200), viewed 30 June 2013, http://www.ofa.org.au/papers/OFA_Newspoll_Report_2008.pdf

Monbiot, G 2000, Organic Farming Will Feed the World, *The Guardian*, 24 August 2000.

Niggli U, Pers Com (2013), Professor Dr Urs Niggli is the executive Director of FiBL and presented this information at the launch of TIPI at Nuremberg, Germany in February, 2013

OTA 2011, U.S. Families' Organic Attitudes and Beliefs Study, Organic Trade Association, 28 Vernon St, Suite 413, Brattleboro VT 05301, USA,

Posner, JL, Baldock, JO, and Hedtcke, JL 2008, Organic and conventional production systems in the Wisconsin integrated cropping systems trials: I. Productivity 1990–2002. *Agronomy Journal*, vol. 100, no. 2, pp. 253-260.

Parrott, N 2002, 'The Real Green Revolution', Greenpeace Environmental Trust, Canonbury Villas, London, ISBN 1 903907 02 0.

Pimentel, D, Hepperly, P, Hanson, J, Douds, D, and Seidel, R, 2005, Environmental, energetic, and economic comparisons of organic and conventional farming systems. *BioScience*, vol. 55, no. 7, pp. 573-582.

Reganold J, Elliott L and Unger Y1987, Long-term effects of organic and conventional farming on soil erosion. *Nature*, vol. 330, no. 6146, pp. 370-372.

Reganold J, Glover J, Andrews P, and Hinman H 2001, Sustainability of three apple production systems. *Nature*, vol. 410, no. 6831, pp. 926-930.

Rodale 2003, Farm Systems Trial, Rodale Institute, 611 Siegfriedale Road, Kutztown, PA 19530-9320, USA, viewed 30 June 2013, http://rodaleinstitute.org/our-work/farming-systems-trial/farming-systems-trial-30-year-report/

Rodale 2006, No-Till Revolution, Rodale Institute, 611 Siegfriedale Road, Kutztown, PA 19530-9320 USA, viewed 30 June 2013 http://rodaleinstitute.org/our-work/organic-no-till/

Seufert V, Ramankutty N and Foley JA 2012, Comparing the yields of organic and conventional agriculture, *Nature,* vol. 485, no. 7397. pp. 229–232.

UNEP-UNCTAD 2008, United National Conference on Trade and Development (UNCTAD) and the United Nations Environment Programme (UNEP), Organic Agriculture and Food Security in Africa, viewed 30 June 2013, http://www.unctad.org/en/docs/ditcted200715_en.pdf

Wang Mau Hua (2012), Pers Com, Director of the Green Food Division of the Ministry of Agriculture quoted these figures at the opening ceremonies of Biofach China in Shanghai, 24-05-2012 and the Organic Trade Union of China Summit in Chengdu China on 17-11-2013.

Welsh R, 1999, Henry A. Wallace Institute, The Economics of Organic Grain and Soybean Production in the Midwestern United States, *Policy Studies Report* No. 13, May 1999.

Willer, H, 2011, *The World of Organic Agriculture 2012*: Summary. the world of organic agriculture, viewed 11 June 2013, the world of organic agriculture, 2011-systems-comparison.fibl.org.

Wynen, E, 2003, Organic Agriculture in Australia – Levies and Expenditure, RIRDC 2003, ISBN:0-642-58570-9. Chapter 8 – References Australian Government 2012, Australia in the Asian Century: White Paper, Australian Government Publishing Service, Canberra.

## Chapter Eight

Brennan, C, Gallagher, K and McEachern, M 2003, A review of the consumer interest in organic meat, *International Journal of Consumer Studies*, vol. 27, no. 5, pp. 381-394.

Chang, H-S and Zepeda, L 2005, Consumer perceptions and demand for organic food in Australia: focus group discussions, *Renewable Agriculture and Food Systems*, vol. 20, no. 3, pp. 155-167.

Chen, J, 2012, A study investigating the determinants of consumer buyer behaviour relating to the purchase of organic food products in urban China. Unpublished PhD thesis, Swinburne University of Technology, Australia.

Chen, J & Lobo, A 2012, Consumption of organic food in urban China: investigating determinants important to buyers and a segmentation analysis of their usage pattern, in Business growth in emerging markets: a debate on critical perspectives, Nova, Hauppauge, New York.

CGFDC 2012, About China Green Food Development Center, viewed 30 March 2012, http://www.greenfood.org.cn/sites/GREENFOOD/List_3675_3811.html.

Euromonitor 2011a, Certified organic: recession no serious threat to organic food and drinks(Part 1), viewed 18 Dec 2012, http://www.euromonitor.com/certified-organic-recession-no-serious-threat-to-organic-food-and-drinks-part-1/report.

Euromonitor International 2011b, Organic packaged food in China, viewed 14 December 2012, http://www.portal.euromonitor.com/Portal/Pages/Analysis/AnalysisPage.aspx.

Fotopoulos, C and Krystallis, A 2002, Organic product avoidance: reasons for rejection and potential buyers' identification in a countrywide survey, *British Food Journal*, vol. 104, nos. 3-5, pp. 233-260.

Gao, Z, Zhao, K, Xiao, X and Tai, C 2009, *Organic agriculture and organic food*, China environmental science publishing House, Beijing (in Chinese).

He, Na, Yang Wanli and Shi, Baoyin 2013, Lost appetite for 'instant chickens', viewed 8 February 2013, http://usa.chinadaily.com.cn/china/201301/07/content_16089880.htm, 2013.

Hsu, JL and Nien, H-P 2008, Who are ethnocentric? Examining consumer ethnocentrism in Chinese societies, *Journal of Consumer Behaviour*, vol. 7, pp. 436-447.

International Trade Centre 2011, *Organic food products in China: market overview*, Geneva, International Trade Centre.

Kluger, J, 2010, What's so great about organic food?, *Time Magazine*, vol. 196, no. 10, 6 September.

Krystallis, A, and Chryssohoidis, G 2005, Consumers' willingness to pay for organic food: factors that affect it and variation per organic product type, *British Food Journal*, vol. 107, no. 5, pp. 320-343.

Li, Y, Cheng, L and Ren, G 2005, Product differences and purchasing decision making on non-environmental pollution agricultural products: An analysis of consumer survey in Jiaxin (in Chinese), *Zhejiang Statistics*, vol. 2, pp. 22-24.

Liu, P 2007, *A Practical Manual for Producers and Exporters from Asia: Regulations, Standards and Certification for Agricultural Exports*, FAO, Rome.

Lobo, A and Chen, J 2013, Marketing of organic food in urban China: an Analysis of consumers' lifestyle segments, *Journal of International Marketing and Exporting*, Vol. 17. no. 1.

Lockie, S, Halpin, D, and Pearson, D 2006, Understanding the market for organic food, in Kristiansen, P, Taji, A, and Reganold, J, (eds), *Organic agriculture: a global perspective*, CSIRO Publishing.

Lockie, S, Lyons, K, Lawrence, G, and Mummery, K 2002, Eating 'green: motivations behind organic food consumption in Australia', *Sociologia Ruralis*, vol. 42, no. 1, pp. 23-40.

Magkos, F, Arvaniti, F and Zampelas, A 2006, Organic food: buying more safety or just peace of mind? A critical review of the literature, *Critical Reviews in Food Science and Nutrition*, vol. 46, no. 1, pp. 23-56.

Organic Food Development Center 2012, OFDC introduction, viewed 28 March 2012, http://www.ofdc.org.cn/about/about.asp.

Pan, Linqing, Wu, Xiaokang and Wei, Shengyao Xinhua News: follow imitated organic food (in Chinese), viewed 1 February 2013, http://news.xinhuanet.com/politics/2011-10/31/c_111135037.htm.

Redruello, Francisco 2013, *Foreign Milk Formula in China: A Passport to Safety?*, viewed 6 June 2013, http://www.portal.euromonitor.com/Portal/Pages/Magazine/IndustryPage. St-Maurice, I, Süssmuth-Dyckerhoff, C and Tsai, H 2008, What's new with the Chinese consumer, *The McKinsey Quarterly*, October.

Shamdasani, P, Chon-Lin, G O, and Richmond, D 1993, Exploring green consumers in an oriental culture: role of personal and marketing mix factors, *Advances in Consumer Research*, vol. 20, no. 1, pp. 488-493.

Sun, X, and Collins, R 2006, Chinese consumer response to imported fruit: Intended uses and their effect on perceived quality, International Journal of Consumer Studies, vol. 30, no. 2, pp. 179-188.

Ureña, F, Bernabéu, R, and Olmeda, M 2008, Women, men and organic food: differences in their attitudes and willingness to pay: a Spanish case study, *International Journal of Consumer Studies*, vol. 32, no. 1, pp. 18-26.

US Department of Agriculture 2008, The People's Republic of China, organic products: Shanghai organic retail market profile. Vol. CH8821, Gain report.

Wang, M 2010, *The supervision and international cooperation of China organic product certification*, BioFach Conference on International Organic Food Markets and Development, China Green Food Development Center, Beijing.

Wang, CL, and Chen, ZX 2004, Consumer ethnocentrism and willingness to buy domestic products in a developing country setting: testing moderating effects, *Journal of Consumer Marketing*, vol. 21, no. 6, pp. 391-400.

Willer, H, and Kilcher, L (eds) 2012, *The world of organic agriculture: statistics and emerging trends 2012*, International Federation of Organic Agriculture Movements, Bonn: Frick: FiBL 2012.

World Bank 2013, GDP per capita, viewed 14 Feb 2013, http://search.worldbank.org/data?qterm-World%20average%20GDP%20per%20cap%3Bita&language=EN.

Xie, W, and Xiao, X 2007, Country Report on Organic Agriculture in China, International Trade Centre's Regional Conference on Organic Agriculture in Asia, December 12-15, 2007, Bangkok, Thailand.

Yan, Y 2012, Food Safety and Social Risk in Contemporary China, *The Journal of Asian Studies*, vol. 71, no. 03, pp. 705-729.

Yang, KCC 2004, A comparison of attitudes towards internet advertising among lifestyle segments in Taiwan, *Journal of Marketing Communications*, vol. 10, no. 3, pp. 195-212.

Yau, OH, Chan, TS and Lau, KF 1999, Influence of Chinese cultural values on consumer behaviour: a proposed model of gift-purchasing behaviour in Hong Kong, *Journal of International Consumer Marketing*, Vol. 7, pp. 97-112.

## Chapter Nine

Abramitzky, R, and Braggion, F 2009, *Malthusian and Neo-Malthusian Theories: Malthus' Legacy*.

ActionAid 2003, *Food Aid: An ActionAid Briefing Paper*, ActionAid.

Andersen, SA 1990, Core Indicators of Nutritional State for Difficult to Sample Populations. *The Journal of Nutrition*, vol. 120 (Suppl 11), pp. 1559-1600.

Balfour, LEB 1975, The living soil and The Haughley experiment, Faber and Faber. London

Biraben, JN 1980, An Essay Concerning Mankind's Evolution. *Journal of Human Evolution*, vol. 9, no. 8, pp. 655-663.

Blakemore, RJ 2000, Ecology of Earthworms under the 'Haughley Experiment' of Organic and Conventional Management Regimes. *Biological Agriculture and Horticulture*, vol. 18, pp. 141-159.

Boserup, E 1965, *The Conditions of Agricultural Growth: The Economics of Agrarian Change under Population Pressure*, G. Allen and Unwin, London.

Budge, T and Slade, C 2009, *Integrating Land Use Planning and Community Food Security. Victoria, Australia*, Victorian Local Governance Association: 76.

Carson, R 1962, *Silent Spring*, New York, Houghton Mifflin.

Chalmers, C 1852, *Notes, thoughts, and inquiries. Princes Street, Soho*: London, John Churchill.

Chappell, MJ and LaValle, LA 2011, Food security and biodiversity: can we have both? An agroecological analysis. *Agriculture and Human Values*, vol. 28, pp. 3-26.

Daily, GC, Ehrlich, AH, and Ehrlich, PR 1994, Optimum human population size, *Population and Environment*, vol. 15, no. 6, pp. 469-475.

Darcy, J and Hofmann, C 2003, *According to Need? Needs assessment and decision making in the humanitarian sector*, Overseas Development Institute, London

Earth Summit 2002, Decision Making: Briefing Sheet. viewed 20 July 2012, http://www.earthsummit2002.org/es/life/Decision-making.pdf.

Ericksen PJ 2008, Conceptualizing food systems for global environmental change research, *Global Environmental Change*, vol. 18, no. 1, pp. 234-245

EuropeAid 2012, About EuropeAid. viewed 13 Feb 2013, http://ec.europa.eu/europeaid/infopoint/publications/europeaid/documents/163a_en.pdf.

FANTA2 2010, About Food and Nutrition Technical Assistance II Project (FANTA-2). viewed 15 July 2012, http://www.fantaproject.org/.

FAO 1946, Report of the First Session of the Conference held at the City of Quebec, Canada, 16 October to 1 November, 1945, Washington, FAO: 89.

FAO 2000, Handbook for Defining and Setting up a Food Security Information and Early Warning System (FSIEWS). ROME, Food and Agriculture Organisation, viewed 22 January 2013, www.fao.org/nr/climpag/pub/Manual%20of%20FSIEWS.pdf

FAO 2002, Food, Nutrition and Agriculture. Rome, FAO Food and Nutrition Division, viewed 26 January 2010, http://www.fao.org/about/en/

FAO 2003, Trade Reforms and Food Security: Conceptualizing the Linkages. Rome, Food and Agriculture of the United Nations, viewed 26 January 2010, http://www.fao.org/about/en/

FAO 2006, Policy Brief: Food Security. Rome, Food and Agriculture Organisation, viewed 26 January 2010.

FAO 2013, The Food and Agriculture Organization of the United Nations. viewed 26 January 2010, http://www.fao.org/about/en/.

FAOSTAT 2013, Food and Agriculture Statistics, Food and Agriculture Organisation, viewed 26 January 2013, www.fao.org/statistics/

FEWSNET 2010, SPECIAL BRIEF: Revising the FEWS NET Food Insecurity

Severity Scale. Washington, US Agency for International Development (USAID).

FIAN 2010, Website of the FoodFirst Information and Action Network. viewed 12 July 2011, http://www.fian.org/

FIVMS 2008, The FIVIMS Initiative: Food Insecurity and Vulnerability Information and Mapping Systems: Tools and Tips, viewed 22 January 2013, www.fao.org/docrep/004/ab990e/ab990e05.htm

Floro, M. and R. Swain 2010, *Food Security, Gender and Occupational Choice among Urban Low-Income Households*. Working Papers, American University, Department of Economics, viewed 22 June 2013, www.american.edu/cas/economics/pdf/upload/2010-6.pdf

Freen, RH 1996, Ancient Greek philosophical concerns with population and environment. *Population and Environment*, vol. 17, no. 6, pp. 447-458.

Gardner, BL 2002, *American Agriculture in the Twentieth Century: How it Flourished and What it Cost*, Massachusetts, Harvard University Press, Cambridge.

Gibson, M 2012, *The Feeding of Nations: Re-defining Food Security for the 21st Century*. Bica Raton, CRC Press Flotida.

Gilland, B 2006, Population, nutrition and agriculture. *Population and Environment*, vol. 28, no.1, pp.1-16.

GMES 2007, *Global Monitoring of Environment and Security*. viewed 12 March 2011, http://www.gmfs.info/.

Greenway, K 2008, *Inter-Agency Task Team on Children and HIV and AIDS: Working Group on Food Security and Nutrition*. Paris, IATT Food Security and Nutrition Working Group.

Grigg, D 1982, *The Dynamics of Agricultural Change: The Historical Experience*, St Martin's Press, New York.

Gustavsson, J, Cederberg, C, Sonesson, U, Van Otterdijk, R, and Meybeck, A 2011, *Global food losses and food waste*. Rome, Italy: Food and Agriculture Organization of the United Nations.

Haub, C 2002, How Many People Have Ever Lived on Earth? *Population Reference Bureau*, vol. 30, no. 8, pp. 3-4.

Holland, BK 1993, A View of Population Growth Circa A.D. 200. *Population and Development Review*, vol. 19, no. 2, pp. 328-329.

IAASTD 2009, IAASTD Global Report: Agriculture at a crossroads. IAASTD Synthesis Report. Johannesburg, South Africa, International Assessment of Agricultural Knowledge, Science and Technology for Development (IAASTD): 6.

IFAD 2009, Food Security: A Conceptual Framework. viewed 15 August 2009, http://www.ifad.org/hfs/thematic/rural/rural_2.htm

James, C 2010, *Global Status of Commercialized Biotech/GM Crops*. 2010. Philippines, The International Service for the Acquisition of Agri-biotech Applications (ISAAA).

Johnson, DG 1997, Agriculture and the Wealth of. Nations. *American Economic Review*, vol. 87, no. 2, pp. 1-12.

Johnson, DG 2000, Population, Food, and Knowledge. *The American Economic Review*, vol.90. no. 1, pp.1-14.

Lidicker Jr, WZ 1962, Emigration as a Possible Mechanism Permitting the Regulation of Population Density Below Carrying Capacity. *The American Naturalist*, vol. 96, no. 886, pp. 29-33.

Malthus, TR 1798, *An Essay on the Principle of Population, As It Affects the Future Improvement of Society with Remarks on the Speculations of Mr. Godwin, M. Condorcet, and Other Writers*. London, Printed for J. Johnson, in St. Paul's Church-Yard.

Malthus, TR 1803, *An Essay on the Principle of Population; or, a View of its Past and Present Effects on Human Happiness; with an enquiry into our Prospects respecting the Future Removal or Mitigation of the Evils which it occasions*. London, Printed for J. Johnson, in St. Paul's Church-Yard.

Maxwell, S, and Frankenberger, T 1992, *Household Food Security: Concepts, Indicators and Measurements: A Technical Review*. New York, Rome, UNICEF and IFAD.

McCalla, AF and Revoredo, CL 2001, *Prospects for Global Food Security: a Critical Appraisal of Past Projections and Predictions*. Washington, DC, International Food Policy Research Institute.

McEvedy, C, and Jones, R 1974, *Atlas of world population history*. New York, Facts on File.

European Commission 2010, Preparatory study on food waste across EU 27. European Commission (DG ENV), Directorate C – Industry.

Nathoo, T. and Shoveller, J 2003, Do healthy food baskets assess food security? *Chronic Diseases in Canada*, vol 24, no. 2-3, pp. 65-69.

Northbourne, L 1940, *Look to the Land*, J. M. Dent and Sons, London

Parfitt, J, Barthel, M, and Macnaughton, S 2010, Food waste within food supply chains: quantification and potential for change to 2050. *Philosophical Transactions of the Royal Society B: Biological Sciences*, vol. 365, no. 1554, pp. 3065-3081.

Riely, F, Mock, N, Cogill, B, Bailey, L, and Kenefick, E 1995. *Food security indicators and framework for use in the monitoring and evaluation of food aid programs*. IMPACT: Food security and nutrition monitoring project, Arlington, Va., U.S.A.

Rijsberman, F 2010, *Energy and Climate: Water and Food Security*. America.Gov March.

Roughgarden, J 1979, *Theory of Population Genetics and Evolutionary Ecology: An Introduction*. New York, Macmillan.

Ruxin, JN 1996, *Hunger, Science, and Politics: FAO, WHO, and Unicef Nutrition Policies, 1945-1978*, Chapter II, The Backdrop of UN Nutrition Agencies, by Joshua Nalibow Ruxin. London, University College London.

Scaramozzino, P 2006, *Measuring Vulnerability to Food Insecurity*. Rome, FAO Agricultural and Development Economics Division (ESA): 26.

Sen, A 1981, *Poverty and Famines: An Essay on Entitlement and Deprivation*, Clarendon Press, Oxford.

SOFI 2001, *The State of Food Insecurity in the World 2001*. Rome, Food and Agriculture Organisation, viewed 22 July 2013, http://orton.catie.ac.cr/cgi-bin/wxis.exe/?IsisScript=LVV.xis andmethod=post andformato=2 andcantidad=1 andexpresion=mfn=002345

SOFI 2011, *The State of Food Insecurity in the World 2011*. SOFI. Rome, Food and Agriculture Organisation, viewed 22 July 2013, http://orton.catie.ac.cr/cgi-bin/wxis.exe/?IsisScript=LVV.xis andmethod=post andformato=2 andcantidad=1 andexpresion=mfn=002345

Stuart, T 2009, *Waste: Uncovering the Global Food Scandal*, W.W. Norton. New York.

Tomlinson, H 1978, 'Not an Instrument of Punishment': Prison Diet in the Mid-Nineteenth Century. *International Journal of Consumer Studies*, vol. 2, no. 1, pp. 15-26.

UN 1973, The Determinants and Consequences of Population Trends: New Summary of Findings on Interaction of Demographic, Economic and Social Factors, vol. I, *Population Studies*, No. 50. Sales No. E.71.XIII.5. United Nations, New York,.

UNDP 2006, *Human Development Report 2006 – Beyond Scarcity: Power, Poverty and the Global Water Crisis*, United Nations Development Programme, New York.

UNPP 2009, *World Population Prospects: The 2008 Revision Population Database*, United Nations Population Fund.

USAID 1992, *Policy Determination: Definition of Food Security*. USAID, US Agency for International Development.

USAID 2007, *Food for Peace*: FY 2008 P.L. 480 Title II Program Policies and Proposal Guidelines, Washington, D.C, United States Agency for International Development.

USAID 2010, *Our Work: A Better Future For All*. US Agency for International Development. viewed 12 June 2011, http://www.usaid.gov/our_work/

USAID 2013, *What is The Famine Early Warning Systems Network* (FEWS NET)?. viewed 30 May 2013, http://www.fews.net/ml/en/info/Pages/default.aspx?l=en

USCB 2008, *Historical Estimates of World Population*. Washington, D.C., U.S. Census Bureau.

USCB 2008, *Total Midyear Population for the World: 1950-2050*, U.S. Census Bureau.

USDA 2009, *Food Security in the United States: Measuring Household Food Security*. viewed 15 October 2009, http://www.ers.usda.gov/Briefing/FoodSecurity/measurement.htm.

USDA/ERS 2010, *Food Security Assessment, 2010-20*. GFA: Food Security Assessment. Washington, Unoted States Department of Agriculture.

Van Den Bergh, JCJM and Rietveld, P 2004, Reconsidering the limits to world population: Meta-analysis and meta-prediction. *Bioscience*, vol. 54, no. 3, pp. 195-204.

Watson, A 1983, *Agricultural Innovation in the Early Islamic World*. U.K., University Press, Cambridge.

Watson, AM 1974, The Arab Agricultural Revolution and Its Diffusion, 700-1100. *The Journal of Economic History*, vol. 34, no. 1, pp. 8-35.

WEF 2009, *Nutrition. Summit on the Global Agenda 2009*, Council Reports 2009.

Weisdorf, JL 2005, From Foraging to Farming: Explaining the Neolithic Revolution, *Journal of Economic Surveys*, vol. 19, no. 4, pp. 561-586.

Woodruff, CE 1909. *The Expansion of the Races*, Rebman Company, New York.

Zhang, W 2008, A forecast analysis on world population and urbanization process. *Environment Development and Sustainability*, vol. 10, no. 6, pp. 717-730.

## Chapter Ten

Altieri, MA 2002, Agroecology: the Science of Natural Resources Management for Poor Farmers in Marginal Environments, *Agriculture, Ecosystems and Environment*, Vol. 93, pp. 1-24.

Altieri, M. and Toledo, V 2011, The Agroecological Revolution in Latin America: Rescuing Nature, Ensuring Food Sovereignty and Empowering Peasants, *The Journal of Peasant Studies*, Vol. 38, no. 3, pp. 687-612.

Allen, P and Hinrichs, C 2007, *Buying into 'Buy Local': Engagements of United States Local Food Initiatives. Constructing alternative food geographies: Representation and practice*, pp. 255-272.

Beck, U 2000, *What is Globalisation?* Polity Press, Cambridge.

Beck, U 1999, *World risk society*, Wiley-Blackwell.

Blay-Palmer, A 2008, *Food Fears: From Industrial to Sustainable Food Systems*, Ashgate Publishing, Aldershot.

Bostrom, M 2005, *Digesting Public Opinion: A Meta-Analysis of Attitudes Toward Food, Health and Farms*, Frameworks Institute.

Bovè, J and Dufour, F 2002, *Food for the Future: Agriculture for a Global Age*, Polity Press. Cambridge.

Davis, A 2009, Cargill's Inside View Helps it Buck Downturn, *The Wall Street Journal*, viewed 12 March, 2012, http://online.wsj.com/article/SB123189501407679581.html,

*Declaration of Nyèlèni, 2007*, Declaration of Nyèlèni, viewed 7 April 2012, http://www.nyeleni.org/spip.php?article290.

Desmarais, AA 2007, La vía campesina. *Wiley-Blackwell Encyclopedia of Globalization*.

Edelman, M 2008, Transnational Organising in Agrarian Central America: Histories, Challenges, Prospects, *Journal of Agrarian Change*, vol. 8, no. 2-3, pp. 229-257.

Escobar, A 1995, *Encountering Development: The Making and Unmaking of the Third World.* Princeton. Princeton University Press, New Jersey.

Friedland, W 2008, *Agency and the agrifood system in The Fight over Food: Producers, consumers, and activists challenge the global food system.* Wright, W. and Middendorf, G. (eds.), 2008, University Park, The Pennsylvania State University Press, Pennsylvania.

Getz, A 1991, Urban Foodshed, *The Permaculture Activist*, October: 26-27.

Goodman, D 2004, Rural Europe Redux? Reflections on Alternative Agro-Food Networks and Paradigm Change, Sociologia Ruralis, Vol. 44, no. 1, pp. 3-16.

Goodman, D and Goodman, M 2007, Localism. Livelihoods and the 'Post Organic': Changing Perspectives on Alternative Food Networks in the United States, D. Maye, L. Holloway and M. Kneafsey (Eds.), *Alternative Food Geographies: Representation and Practices*, Oxford, Elsevier, pp. 23-38.

Goodman, D and Goodman, M 2001, Sustaining Foods: Organic Consumption and the Socio-Ecological Imaginary, in Choen, M. and Murphy, J. (eds) *Exploring Sustainable Consumption: Environmental Policy and the Social Sciences*, Amsterdam, Pergamon, pp. 97-19.

Gottlieb, R, and Fisher, A 1996, Community Food Security and Environmental Justice: Searching for a Common Discourse. *Agriculture and Human Values*, Vol. 3, no 3, pp. 23-32.

Guthman, J 2007, Commentary on Teaching Food: Why I am Fed Up with Michael Pollan, *Agriculture and Human Values*, Vol. 24, pp. 261-264.

Hardt, M and Negri, A 2000, Empire, Harvard University Press, Cambridge.

Harvey, D 1996, Justice, *Nature and the Geography of Difference*, Wiley-Blackwell, Cambridge.

Hendrickson, MK, and Heffernan, WD 2002, Opening Spaces through Relocalisation: Locating Potential Resistance in the Weaknesses of the Global Food System. *Sociologia Ruralis*, vol. 42, no. 4, pp. 347-369.

Hendrickson, MK, Wilkinson, J., Heffernan, WD, and Gronski, R 2008, *The*

*Global Food System and Nodes of Power*, viewed 12 December 2012, http://ssrn.com/abstract=1337273.

Hendrix, CS 2011, *Markets vs. Malthus: Food Security and the Global Economy*, Policy Brief, Peterson Institute for International Economics.

Holt-Gimènez, E 2010, Food Security, Food Justice, or Food Sovereignty?, *Food First Backgrounder*, vol. 16, no. 4, pp. 1-4.

Holt-Gimènez, E and Altieri, M 2013, Agroecology, Food Sovereignty, and the New Green Revolution, *Agroecology and Sustainable Food Systems*, vol. 37, no. 1, pp. 90-102.

Holt-Gimènez, E and Shattuck, A 2011, Food Crises, Food Regimes and Food Movements: Rumblings of Reform or Tides of Transformation? *Journal of Peasant Studies*, Vol. 38, no. 1, pp. 109-144.

IAASTD 2009, *International Assessment of Agricultural Knowledge, Science and Technology for Development*, viewed 10 February, 2013, http://www.agassessment.org/.

Johnston, J, Biro, A and MacKendrick, N 2009, Lost in the supermarket: the corporate-organic foodscape and the struggle for food democracy, *Antipode*, vol. 41, no. 3, pp. 509-532.

Kerssen, T 2012, Hunger is Political: Food Sovereignty Prize Honours Social Movements, *Food First*, viewed 12 March, 2012, http://www.foodfirst.org/en/node/4020.

Kerssen, T 2009 Gates agriculture speech highlights sustainability but falls short, *Food First*, viewed 17 December, 2009, http://www.foodfirst.org/en/node/2608.

Kjaernes, U Harvey, M and Warde, A 2007, *Trust in Food*, Macmillan, New York.

Kloppenburg, JJ, Hendrickson, J., and Stevenson, GW 1996, Coming in to the Foodshed, *Agriculture and Human Values*, vol. 13, no. 3, pp. 33-42.

Kneen, B 2011, *Invisible Giants, Cargill and its Transnational Strategies*, Macmillan, New York.

Lang, T 2003, Food Industrialisation and Food Power: Implications for Food Governance. *Development Policy Review*, vol. 21, no. 5-6, pp. 555–568.

Lang, T, and Heasman, M 2004, Diet and Nutrition Policy: A clash of ideas or investment? *Development*, vol. 47, no. 2. pp. 64-74.

La Via Campesina, 2009a, *La Via Campesina Policy Documents: 5th Conference*, viewed January 20, 2011, http://viacampesina.org/downloads/pdf/policydocuments-/POLICYDOCUMENTS-EN-FINAL.pdf .

La Via Campesina, 2009b. *G8-Farming: The G8 should clean up their own mess*, viewed 17 April, 2009, http://alainet.org/active/29971.

La Via Campesina, 2007, *The Nyèlèni Declaration* http://www.nyeleni.org/spip.php?article290, accessed 10 June, 2010.

La Via Campesina, 2006, *Neoliberal policies are wrecking food sovereignty*, viewed 12 March 2013, http://viacampesina.org/en/index.php/main-issues-mainmenu-27/food-sovereignty-and-trade-mainmenu-38/33-food-sovereignty.

La Via Campesina, 1996, *The right to produce and access to land. Voice of the Turtle.* http://www.voiceoftheturtle.org/library/1996%20Declaration%20of%20Food%20Sovereignty.pdf, ccessed 14 March, 2009.

Marsden, T 1997, Creating Space for Food: The Distinctiveness of Recent Agrarian Development. In D. Goodman and M. Watts (Eds.), *Globalising Food: Agrarian Questions and Global Restructuring*, London: Routledge.

McMichael, P 2011, *The Food Regime in the Land Grab*, Paper presented at the International Conference on Global Land Grabbing, April 10-11, University of Sussex, viewed 15 April, 2013, http://www.iss.nl/fileadmin/ASSETS/iss/Documents/Conference_papers-/LDPI/77_Philip_McMichael.pdf.

McMichael, P 2008, Peasants Make Their Own History, But Not Just as They Please, *Journal of Agrarian Change*. vol. 8, no. 2-3, pp. 205-228.

McMichael, P 2004, *Global Development and the Corporate Food Regime*, Prepared for Symposium on New Directions in the Sociology of Global Development, XI World Congress of Rural Sociology, Trodheim. July 2004, viewed 12 March 2012, http://www.infoagro.net/shared/docs/a1/Global%20development%20and%20the%20corporate%20food%20regime.pdf.

McMichael, P 2000, The Power of Food, *Agriculture and Human Values*, vol. 17, pp. 21-33.

McMichael, P and Friedmann, H 2007, Situating the 'Retailing Revolution'. Supermarkets and agri-food supply chains: Transformations in the production and consumption of foods, 291.

Meyer, G 2010. Bunge rides on volatility of food markets, *Financial Times*, viewed 20 March 2013, http://www.ft.com/intl/cms/s/0/89e80c8a-12a8-11e0-b4c8-00144feabdc0.html#axzz2cfcMDTXg.

Micheletti, M 2003, *Political Virtue and Shopping: Individuals, Consumerism, and Collective Action*, Palgrave Macmillan, New York.

Mulvany, P 2008, *Food at Any Price is Not Sustainable, Food Ethics Council*, Vol. 3, no. 2, viewed 12 March, 2013, http://www.foodethicscouncil.org.

Murdoch, J, Marsden, T., and Banks, J. 2000, Quality, Nature, and Embeddedness: Some Theoretical Considerations in the Context of the Food Sector, *Economic Geography*, Vol. 76, no. 2, pp. 107-125.

Murphy, S, Burch, D. and Clapp, J 2012, *Cereal Secrets: The World's Largest Grain traders and Global Agriculture*, Oxfam International, viewed 14 February, 2012, http://www.-oxfam.org/en/grow/policy/cereal-secrets-worlds-largest-grain-traders-global-agriculture.

Patel, R 2009, *Stuffed and Starved: markets, Power and the hidden Battle for the World Food System*, Black Inc., Melbourne.

Patel, R 2006, International Agrarian Restructuring and the Practical Ethics of Peasant Movement Solidarity, *Journal of Asian and African Studies*, vol.41, no.1/2, pp.71-93.

Patel, R, Balakrishnan, R., and Narayan, U 2007, Transgressing rights: La Via Campesina's call for food sovereignty/Exploring collaborations: Heterodox economics and an economic social rights framework/Workers in the informal sector: Special challenges for economic human rights. *Feminist Economics*, vol. 13, no. 1, pp. 87-116.

Pollan, M 2001, *The Botany of Desire: A Plant's Eye View of the World*, Random House, New York.

Petrini, C. and Padovani, G 2006, *Slow Food Revolution: A New Culture for Eating and Living*, Rizzoli, New York.

Scott, JC 1985, *Weapons of the Weak: Everyday Forms of Peasant Resistance*, University Press, New Haven, Yale.

Slater, D 2004, *Geopolitics and the Post-Colonial: Rethinking North-South Relations*, Blackwell Publishing, Oxford.

*Surin Declaration*, 2012, Surin Declaration: First Global Encounter on Agroecology and Seeds, viewed 17 March, 2013, http://viacampesina.org/en/index.php/main-issues-mainmenu-27/sustainable-peasants-agriculture-mainmenu-42/1334-surin-declaration-first-global-encounter-on-agroecology-and-peasant-seeds.

Padel, S, Röcklinsberg, H and Schmid, O 2009, The Implementation of Organic Principles and Values I the European Regulation for Organic Food, *Food Policy*, Vol. 34, pp. 245-251.

Reitan, R 2007, *Global Activism*, Routledge, New York.

Robin, M. and Holoch, G 2010, *The World According to Monsanto: Pollution, Politics and Power*, Vic, Spinifex Press, North Melbourne.

Rosset, P 2006, Moving Forward: Agrarian Reform as Part of Food Sovereignty, in Rosset, P., Patel, R., and Courville, M. 2006, *Promised Land: Competing Visions of Agricultural Reform*, Oakland, California, Food First Books, pp. 301-322.

Saragih, H., 2008, *Open Letter. Via Campesina Proposal to Solve Food Crisis: Strengthening Peasant and Farmer-based Food Production*, viewed 11 April, 2010, http://www.-foodfirst.org/en/node/2109.

Seyfang, G 2006, Ecological citizenship and sustainable consumption: Examining local organic food networks. *Journal of Rural Studies*, vol. 22, no. 4, pp. 383-395.

Van Der Ploeg, J 2008, *The New Peasantries: Struggles for Autonomy and Sustainability in an Era of Empire and Globalisation*, Earthscan, Oxford.

Van Der Ploeg, J., n.d., *The Peasant Mode of Production Revisted*, viewed 17 March 2013, http://www.jandouwevanderploeg.com/EN/publications/articles/the-peasant-mode-of-production-revisited/.

Watts, DCH., Ilbery, B, and Maye, D 2005, Making Reconnections in Agro-food Geography: Alternative Systems of Food Provision, *Progress in Human Geography*, Vol. 29, No. 1, pp. 22-40.

Weis, T 2007, *The Global Food Economy: The Battle for the Future of Farming*. Zed Books, London.

Wittman, H 2009, Reworking the Metabolic Rift: La Via Campesina, Agrarian

Citizenship, and Food Sovereignty, *Journal of Peasant Studies*, vol. 36, no. 4, pp. 805-826.

World Food Program 2010, Who Are the Hungry? viewed 15 March, 2012, http://www.wfp.org/hunger/who-are.

# Biographies of authors

**Dr. Jue Chen** is of Chinese origin and currently a lecturer in the Faculty of Business and Enterprise, Swinburne University of Technology, Australia. Her PhD thesis was titled 'A study investigating the determinants of consumer buyer behaviour relating to the purchase of organic food products in urban China'. This was the first national survey of organic food consumer buyer behaviour in China. Before becoming an academic, she was a director and manager across several businesses, enabling her to acquire an appreciation of global business in its various facets. Jue's special research interests spans across food safety and environmental issues, organic and healthy food, with a particular focus on China.

**Liz Clay** is a certified organic farmer of 22 years producing a range of vegetables, berries and beef for local markets. She is a strong advocate for the environment and sustainable agriculture and has been actively involved in the development of the organic sector at a local, national and international level. In 1998 she became the first Australian to be elected to World Board of the global peak body for the organic industry – the International Federation of Organic Agriculture Movements (IFOAM) and served on that Board for seven years. Closer to home she chaired the Victorian Organic Industry Committee and was appointed as Chair of the West Gippsland Catchment Management Authority from 2003 to 2011.

**Dr. Mark Gibson** worked in the food industry for over two decades before laying down his knives to pursue a career in academia. After dealing with food at an artistic level for so long Mark became interested in the wider aspects of food, particularly of food and culture. Taking this interest further Mark pursued this topic in his PhD degree. While he remains passionate about all things gastronomic. Mark's focus has shifted from foodservice to all areas of social and environmental issues concerning food. His interests are particularly diverse and include such areas as sustainability; of land and water usage, pollution, climate change, population carrying capacity, poverty, rights and equality issues, food and social conflict, environmental politics and of food production. Mark now studies and teaches in Asia.

**Dr. Chandana Hewege** is a lecturer in International Business at the Faculty of Business and Enterprise, Swinburne University of Technology. He has earned a PhD in International business from the Monash University (Australia), an MBA and a Bachelor's degree in Business Administration from the University of Sri Jayewardenepura (Sri Lanka), a diploma in international SME management from the Linkoping University (Sweden), an advanced certificate in Marketing from the Chartered Institute of Marketing (UK) and a Licentiate certificate in Accounting from the Institute of Chartered Accountants in Sri Lanka. Chandana's academic and professional experiences are predominantly centered on teaching, research and industry consultancy. He has published about 34 refereed research and conference papers. He possesses about 15 years of teaching experience at both undergraduate and postgraduate levels, local as well as overseas.

**Andre Leu** is the President of International Federation of Organic Agricultural Movements (IFOAM), the world umbrella body for the organic sector. Andre has 40 years of experience in all areas of organic agriculture from growing, pest control, weed management, marketing, post-harvest, transport, grower organisations and developing new crops and related education in Australia and in many other countries. He has an extensive knowledge of farming and environmental systems across Asia, Europe, the Americas, and Africa from over 35 years of visiting and working these countries. He has written and published extensively in magazines, newspapers, journals, conference proceeding, newsletters, website and other media on many areas of organic agriculture. Andre and his wife own an organic tropical fruit orchard, in Daintree, Queensland, Australia that supplied quality controlled fruit to a range of markets from local to international.

**Associate Professor Antonio Lobo** teaches Marketing and Supply Chain Management to postgraduate students in the Faculty of Business and Enterprise at the Swinburne University of Technology, Melbourne, Australia. Previously he has taught in Malaysia, Singapore and Tasmania. His research interests include supply chain management, services marketing and consumer behaviour. Antonio's research has been published in the Journal of Consumer Marketing, International Journal of Value Chain Management, Asia Pacific Journal of Marketing and Logistics, Services Marketing Quarterly and Journal of Travel and Tourism Marketing among others. He has been a recipient of several grants for research in organic food and has successfully supervised several PhD students in marketing and supply chain management.

**Dr Alana Mann** is a lecturer and researcher in the Media and Communications Department at the University of Sydney. Her book Power Shift: Global Activism in Food Politics (2013) focuses on the international farmers' movement La Via Campesina and campaigns for food sovereignty in Chile, Mexico and Spain.

**Associate Professor Bruno Mascitelli** is an Associate Professor in International Studies at the Swinburne University of Technology, Melbourne, Australia. Prior to joining Swinburne University, Bruno was employed by the Australian Trade Commission for 18 years in the promotion of Australian exports. He has published nine books in areas including Australian and international business, European studies, migration and expatriate voting. He has been a recipient of several grants for research in organic food and especially in areas such as food security and China.

**Dr. Andrew Monk** has two decades of experience in the organic industry auditing, certification and standards, and commercial interests across the organic supply chain, including in horticulture and value adding. Andrew consults to public and private entities across the supply chain on environmental (including organic) issues and management systems.

He is managing director of an environmental sector services company, Mulching Technologies Pty Ltd, Chairman of Biological Farmers of Australia Ltd and an adjunct associate professor in the Australian Centre for Agricultural Law, University of New England, NSW.

**Professor Barry O'Mahony** is Professor of Services Management and Chair of the Department of Marketing, Tourism and Social Investment at Swinburne University of Technology. Barry has taught undergraduate, postgraduate and doctoral courses in Australia, Ireland, Hong Kong, Malaysia and the United States and has developed and delivered undergraduate and postgraduate programmes in hospitality and tourism and food and beverage management. Barry, is a company director of the accrediting body The International Centre of Excellence in Tourism and Hospitality Education, and a member of the academic boards of Le Cordon Bleu and William Angliss Institute.

**Dr. John Paull** is a social scientist and is the editor-in-chief of the open-access peer-reviewed *Journal of Organic Systems*. He has recently been a visiting academic at the University of Oxford. He has degrees in mathematics, psychology and environmental management. He has extensive experience in academic research, corporate research and management, training, and education. He has presented his research at international conferences including Biofach, ISOFAR, and the World Organic Congress. He has published extensively in international journals with some of his papers available at http://orgprints.org. He is the author of books including The Value of Eco-labeling (2009) and book chapters in Marketing of Organic Products (2008), Island Futures (2011) and Diversifying Food and Diets (2013). He is currently at the University of Tasmania and welcomes email contact at j.paull@utas.edu.au.

# INDEX

ABS (Australian Bureau of Statistics), 72-73

ACO (Australian Certified Organics), 58

Agricultural Experimental Circle (AEC), 38, 40-42, 45,

Argentina, 6, 34, 126, 128, 196

AROS (Asian Regional Organic Standard), 130

ASEAN, 129-130

Australia, xiii, xiv-xvi, 2, 6, 8-11, 13-14, 16-18, 21, 32, 34, 37-79, 81, 83, 86, 89, 91, 96, 99-100, 105-107, 110-117, 121-122, 128, 140, 157, 171, 173

Balfour (Lady), 52, 195, 197-198

BFA (Biological Farmers of Australia, 47, 57-58, 66

Brazil, 34, 124, 126, 131, 196, 226

Canada, 34, 63, 75, 122, 126, 128, 206, 222

CERES (Centre for Education and Research in Environmental Strategies), 109

China, vii, 2, 15, 34, 123, 126, 128, 155-173

CSIRO, 52, 60

Dangour study, 23-24, 82

democracy (food), 221, 228-229

European Union, 8, 39, 44, 55, 57, 74, 76, 82, 86, 102, 108

FAO (United Food and Agriculture Organisation), 8, 127, 134, 137, 179, 180, 182, 184-186, 191, 198, 209

first wave (organic production in Australia), 38, 40, 45, 61

Food Miles, 9, 21, 91, 93-94, 110-112, 223

food security, 14-16, 110, 121, 134, 137, 140, 156, 175-198

France, 40, 103

fourth wave, 39, 46, 57-58, 61

"Frankencorn", 35

FTA (Free Trade Agreements), 205, 208

Germany, 4, 9, 40, 48, 83, 86, 103, 113, 153

GMO, iv, 8, 60, 195, 196, 206, 219, 225

GOMA, 127, 130

green food, 123, 156-157, 160-162, 164, 167, 172

green revolution, 1, 147, 156, 201, 204-205, 228

Howard Albert (so-called father of organic food), 2-5, 195

IAEA (International Atomic Energy Agency), 33

IFOAM, xi, xii, xiv, 5-6, 8-9, 26, 39, 46-48, 54, 59-60, 75, 120-121, 124-125, 127-132, 147-148, 161, 202, 225

India, 34, 123-124, 126, 128, 131, 136, 147

Italy, 9, 40-41, 82, 86, 223

Japan, 75, 89, 126, 132-133, 156, 171

*Journal of Organic Systems*, 46, 60

Korea, 47, 63, 75, 126, 133, 151

La Via Campesina, 201, 203-209, 212-216, 223, 226-227

Living Soil Association of Tasmania, (LSAT), 45, 49-51

LOHAS, 87, 89, 91, 93, 95-6, 122, 169

NASAA, (National Association of Sustainable Agriculture in Australia), 46, 57-59, 66

New Zealand, 44, 48, 52, 54, 60, 128, 150

NGO (Non-government organisation), 21, 179, 185, 207, 222, 228

Northbourne Walter, 4, 44-45, 47-48, 195, 197

OECD (Organisation for Economic Cooperation and Development), 18

Organic Food Movement, 46, 53, 54, 201-202

RFID (Radio Frequency Identification), 110, 113-115, 117

second wave (of organic production in Australia), 38, 44-45, 61

Shoobridge Henry, 47-52

*Silent Spring*, 5, 38, 46, 53, 53, 121-122, 195

Soil Association, 5, 23, 38, 45-47, 49-54, 57

Steiner Rudolf, 40-41, 43-45, 47, 61, 137, 194, 197

supply chain, 15, 67-68, 71, 97-117, 159, 171, 203, 216-218, 220, 223

The Australian Organic Farming and Gardening Society (AOFGS), 38, 45, 47-52

third wave (of organic production in Australia), 38, 46

UK (United Kingdom), 5, 22,-23, 33, 38-39, 45, 47-49, 51-55, 57, 68, 86, 103, 108, 141, 190

UN (United Nations), 179, 184, 190-191

UNCTAD, 127, 136-137, 147-148

USA, 4-6, 9, 24, 48, 80, 102, 105, 122, 126, 128, 146, 151, 196,

USA (Department of Agriculture), 21, 23, 158, 184

USAID, 177, 182-186

World Health Organisation (WHO), 27, 160

www.ingramcontent.com/pod-product-compliance
Ingram Content Group UK Ltd.
Pitfield, Milton Keynes, MK11 3LW, UK
UKHW021301180426
11947UKWH00015B/948